RISE ABOVE

Douglas Ell
and
Steve Eggleston

Copyright © 2023 Douglas Ell

All rights reserved. No part of this book may be reproduced, stored in any retrieval system, or transmitted in any form or by any means, electronic, mechanical, photocopying, recording, or otherwise, without express written permission of the author or publisher.

The moral right of the author has been asserted.
First published on the Internet (and England)

ISBN: 9798866525317

Praise for *RISE ABOVE*

An elegantly written book full of imagery, truth, and facts. It is the Gospel told in a fresh and new way to help the lost generation. Wow! What a piece of literature for anyone reflecting on the relation of science, faith, and the material world.

THE RT. REV. PATRICK P. AUGUSTINE,
Episcopal Bishop of South Sudan

From the outset, "Rise Above" promises to bridge the often tumultuous divide between science and spirituality. For a self-professed agnostic like me, the endeavor seemed ambitious, if not audacious. However, the author's meticulous research and balanced approach lend a surprising weight to the argument.

MIKE K.

I've never read any work like it. A witty, humorous, and refreshingly candid saga of a gritty family unfamiliar to biblical faith working through death, doubt, pride, and prejudices. "Rise Above" is an extremely captivating tale that ingeniously folds fascinating character development into an intriguing plot that leads readers to seriously consider the Bible as the Word of God.

DR. RANDY GULIUZZA, *PhD,*
President, Institute for Creation Research

Douglas Ell serves his readers a wealth of historical and scientific information. As the terminally ill protagonist encounters and grapples with scientific evidence for God's existence, he must choose how the final threads of his life will be woven. This novel is a "must read" for all who seek to rise above the cacophony of modernity to find a greater existential purpose.

SCOTT STRIPLING, *PhD, President,*
Near East Archaeological Society

Doug Ell has accomplished something quite extraordinary in creating an engaging work of fiction infused with incisive scientific insights. We are urged to discern the devastating lies and paradigms that hold us captive in spiritual darkness.

SCOTT LANSER, *President,*
Associates for Biblical Research

With Rise Above, Doug Ell has once again produced a culturally challenging albeit scripturally and historically sound text, this time in an approachable and easy to digest novel form. I wholeheartedly recommend this book for anyone open to the truth.

THE VERY REV. FR. JASON A. MURBARGER

Using fictive characters and settings, Doug Ell offers an engaging, well-argued case for embracing God-based science and God-witnessing lives lived out within a universe delightfully designed for human communion with God and neighbor alike.

TIM JORGENSON, *Author of Bonhoeffer;*
The Night Is Far Gone; Advent House; and other titles

Every day I'm ascending the steps of Ivy-covered halls on the campuses of Cambridge Massachusetts, and its distinguished institutions of the highest caliber, and I wonder how its leaders will take the news that the science and religion that they grew up with and went to grad school with isn't what they thought? Try this book.

REV. DAVID R. THOM,
a Chaplain to faculty at Harvard and MIT.

Every family has its challenges. Doug's 'narrative apologetic' approach presents the theological truths of the Christian faith in the lives of people and their stories. I commend Rise Above for its narrative flow and its theological substance and as an illustration of how to share the testimony for the hope that is in you (1 Peter 3:15).

THE REV. FR. CHRISTOPHER M. RODRIGUEZ

RISE ABOVE

What Science Tells Us
About the Existence of God

Douglas Ell

and

Steve Eggleston

"For the wrath of God is revealed from heaven against all ungodliness and unrighteousness of men, who by their unrighteousness suppress the truth. For what can be known about God is plain to them, because God has shown it to them. For his invisible attributes, namely, his eternal power and divine nature, have been clearly perceived, ever since the creation of the world, in the things that have been made. So they are without excuse."

Romans 1:18–20

I wrote this book for you. I pray it will help you rise above the lies of the Devil that have swallowed our culture, help you see beyond the blindness of many of our academic institutions, and free you of the doubts and confusions of our age. For respect and reverence for the LORD are the beginning of wisdom. Proverbs 9:10.

- Douglas Ell

CONTENTS

CHAPTER ONE

It was grey in New York City, but John's mood was black. He staggered out of the cancer center at a quarter to three in the afternoon, Fifth Avenue before him. 99th Street would have been there, but it dead-ended at Madison Avenue into the massive Mount Sinai medical complex.

He saw Luke. Luke wore a black suit and stood outside a black Cadillac Escalade. At a quick glance, it looked like an Uber pickup—nothing special, nothing to see. No one would know Luke was a former Navy SEAL and a bodyguard. No one would suspect the Escalade was heavily armored, with BR7 bulletproof glass, capable of stopping high-velocity assault weapons and sniper bullets, and steel panels to survive grenade attacks. And almost certainly, nobody would recognize John: the thin man in jeans, outdated winter jacket, sunglasses, and simple grey wool cap. John took a perverse pride in dressing shabby on the sidewalks of New York, and, at a quick glance, you might even wonder if he had a roof over his head.

But John was unsteady. He fell. Luke saw and ran over. "You okay, Mr. Arnold?" There was no response.

Luke took John's arm, holding it tightly, and guided him into the back of the car.

"Where to, Mr. Arnold?" There was no response.

Luke started the car and pulled into traffic. It was December 23, and forty blocks ahead was the nightmare of midtown Manhattan at Christmas. Luke took a quick right on the 97th Street transverse and headed west, across Central Park. He would drive until John decided on a destination. The condo on Central Park West? The house in Connecticut?

Luke checked the rearview mirror. John was shaking, his face covered in sweat. Luke saw John take a bottle of pills out of his pocket.

***

John Arnold was more than rich. People knew he was rich, but not how rich. That was how he wanted it, for he was a private person and, at the core of his being, an ordinary man of ordinary tastes.

He grew up poor and didn't go to college. But he had a gift, and what a gift it was. John Arnold was a wizard of the financial world. The people who knew him—and he did his best to keep his name out of the papers and his actions hidden behind corporations and partnerships— were in awe. He could analyze a deal like no other. He could read through a stack of documents and financial tables and know, just know as if he had a sixth sense, where the traps were: the sneaky clauses the lawyers had added, the accounting gaps and deliberately obtuse footnotes, and what was not being said. He could ask the right questions, and he seemed to know, just *know*, the companies and ideas that would change the world.

How does a turtle navigate the open seas and return years later to the same beach? How does a bird take off from Alaska and fly nonstop to Tasmania? That was how John Arnold knew how to make money.

His hedge funds were legendary, his deal-making admired and feared, his gift for spotting a winner inexplicable. And he did it all honestly. He paid his taxes and tried to comply with the inanity of government regulation.

And now, as he sat in the back of his armored Escalade, the verdict was in. He had won the booby prize. "For what does it profit a man to gain the whole world and forfeit his soul?" [1] His kids were lost, distant, or both. One was a criminal and an addict. Two were "successful" (with the help of his money) but kept their distance. He was blessed with a good wife, Saint Mary, who softened his rough edges, his moods and tirades, his grumpy and abrasive nature. He had no close friends—but many pretenders.

He had conquered the material world, charged straight on and beat it in every way possible. He'd fought off financial crises, the treachery of politicians and corporate boards, even those idiot lawyers, but for what? Death, the final enemy, was near. The shadow of death was upon him. He was sweating and breathing heavily. He was in agony, physically and spiritually.

He stared at the bottle of pills. It was small but full: forty pills in total. The pain was bad. His mood was worse. His life was over, almost, and all he had accomplished now seemed rubbish.

And he couldn't remember his dosage. What did the doctors say? Did they give him one pain pill or two? And how long was he supposed to wait before taking more?

Dark thoughts came. They whispered, "Take them all! Be done!" Yes, forty pills. One massive overdose and he would be done. It would be so easy, there was a bottle of water next to his seat. But should he? The thoughts came again, "Take the pills; be done!" He wiped sweat from his forehead and took a deep breath, then another. He pressed down on the lid to twist it open, but his fingers slipped. Anger, frustration, despair, and pain were all he could feel. He was in agony; he was beaten.

"Are you okay, Mr. Arnold? Where do you want to go?"

Luke's questions brought him back. He took a deep breath and was slow to answer.

"I'm not well. I need to think. Find a place to pull over."

Luke took a right on Amsterdam Avenue and headed north. This part of Manhattan was less crowded. A dozen blocks later a parking space opened up, and Luke backed in.

John stared again at the bottle of pills. He wiped his hands on his jeans and tried again. This time, the bottle opened. Now what? One? Two? Forty?

***

There are things that cannot be explained. Something forced John to look up and to his right. It was as if a large invisible hand took a firm hold and rotated his head. On his immediate right was a grassy park and a 40-foot statue.

The Archangel Michael was at the top, his sword piercing Satan. To the north of the fountain, towering over him, staring him in the face as he looked out the window of the Escalade, was the largest church in North America.

He asked Luke. "Why did you pick this spot?" "We got lucky. A car pulled out."

John hummed. "Do you know what that is?"

"No." Luke had his eyes trained on the structure. "Why is a church that big in this part of Manhattan?"

"That's a long story." "How big is that thing?" "Over 600 feet long." "It's lopsided."

"They're still building it. They're further along on the south tower, the one closest to us, here on the west side of the church." John managed a sad smile. "They've been building it since 1892. See that dome on the other side, toward the east?"

"Yeah?" Luke looked to where John was pointing. "That was supposed to be temporary. It was supposed to be a tower 245 feet high. But the ground is unstable; they can't build it. Then, not long ago, it was damaged by fire." John paused.

"Some say it sums up God's church in America. Unfinished, on shaky ground, abandoned by many. They don't have the money to finish it. It's the Cathedral of Saint John the Divine, my namesake so to speak, named after the youngest apostle, who wrote five books of the Bible. He was the only apostle who wasn't killed; he was exiled to an island off Turkey. There, he wrote the book of Revelation, the last book of the Bible, the book of future history, of the Apocalypse to come. He foretold that Jesus will return for the final conquest of evil. That statue on our right is the Fountain of Peace, but there's no water. That's the Archangel Michael, piercing the heart of Satan, in the ultimate triumph of good over evil."

Luke caught his eye in the rearview mirror. "You sure know a lot about this church, Mr. Arnold."

Waves of pain washed over John. He was dizzy, and he paused to catch his breath.

"They asked me for money to finish it." John saw the question in Luke's eyes. "I turned them down."

John bent his head and looked at the open bottle of pills.

CHAPTER TWO

Who knows where it came from? Who knows why it was answered? But from somewhere within John came first a total surrender to the hopelessness of his human condition, and then an absolute and desperate cry for mercy—and forgiveness. And from somewhere in the mystery of existence a force grabbed John and took him on a journey of indescribable terror and wonder. There were colors vivid beyond imagination, colors that made all earthly colors seem pale. There were sounds of soul-shaking depth and harmonies of unimaginable beauty. He heard unspeakable words, words it is not lawful for a human being to utter. [2] Who knows how long it lasted, for it was not of this world, and not measured by human time.

Then it was over. John's earthly eyes reopened. He was different; he was changed. Yet, the outside world remained the same. Luke was sitting in the front seat, the Cathedral before him, the fountain to his right. But he was changed. Now, it wasn't about him. He had seen something greater.

He turned his head slowly from side to side. Was it the cancer? Had he gone crazy?

But he no longer wanted a pill. He was embarrassed he had considered, even for a moment, leaving without saying goodbye to Mary. Slowly, as if handling a live grenade, he screwed the lid back on the bottle of pills, and put it back in his pocket.

"Take me home, Luke. To Connecticut."

Luke drove them out, through the labyrinth of parkways that leads to Connecticut. The world outside flashed by, and the car's rhythm lulled John and calmed his mind. He relaxed, in a way he hadn't truly relaxed in years, as he let go of his need to succeed. Suddenly, unaccountably, it didn't matter, at least for the moment, what the doctors said. It didn't matter that his life was over. It didn't matter that this early afternoon ride was the last in a lifetime of commuting, the last of 10,000 journeys. For some reason, a reason he didn't know, a reason beyond understanding, he began to feel peace. He began to rise above the pain as though he was looking down at himself, and he felt pity for that broken human being.

John was feeling better. He changed his mind. "Take me to the Westchester garage. I'm going to drive myself home."

John kept a garage in Westchester for toys. He didn't need it, and he often felt stupid he had it. His son Mark was always telling him to buy this or that car, and John usually did so to please Mark and give them something to talk about, which was much needed because of their strained relationship. Over the years John had acquired a collection of fantasy cars. He had a ludicrous collection of wines for the same reason. His son Mark wanted him to get them, and he was desperate for a relationship with his son.

Luke unlocked the garage, and John climbed the viewing platform. Facing him were twelve cars, each absurdly expensive, each polished beyond perfection. He chose the red custom Lamborghini. Luke helped him down the platform and into the car. John backed the car out slowly. He hadn't driven the Lamborghini in months, and the power of its engine always scared him. He reacquainted himself with the car's features as he waited for Luke to lock up and start the Escalade. Luke had insisted on following him home, just to be safe.

John's pain and confusion came back. The Lamborghini was ridiculous. Where before he'd felt pride and a touch of snobbery, now he was embarrassed and ashamed. The car was difficult to get into and hurt his back. Who needs a car that can go 220 miles an hour? People were staring, stares that used to make him feel superior. Now, they were laughing at a sad old man in a stupid car. He felt disconnected, as though he was hovering above, watching that sad man drive home.

The satellite radio was tuned to the Holiday Channel, with the usual corny and schmaltzy mix. Christmas was two days away. He couldn't help smiling when he heard "The Chipmunk Song"—surely one of the most annoying Christmas tunes ever. It reminded him of when the kids were small, of "family night" games and pretend wrestling on the carpet. But when George Michael's "Last Christmas" came on, he hit the classical button. How could anyone like a depressing song about getting dumped that bore no connection to a sacred holiday? To his relief, the classical channel was playing the heavenly sounds of Panis Angelicus. Charlotte Church's angelic voice flowed from the car's twenty-four speakers like a balm to his soul.

The nearer to home he got, the bigger, grander, and more pretentious the houses became. Modest colonials faded to larger colonials, and then to monuments to ego. Fake French chateaus and single-family hotels, with acreage. Swimming pools, tennis courts, guest houses. Everything money could buy. For the first time in years, he was astonished by how heavily moneyed the town was. It was like he was seeing it through new eyes.

He pulled off the Merritt Parkway and drove by the neighborhood church, picture perfect with a stone exterior, exquisite stained-glass windows, and slate roof. It used to be plain, almost an embarrassment, until, at Mary's request, he'd paid for a new building. John Arnold was not a man who sat passively on committees. He took over, raised millions (most of it his own money), and made the church a showplace. They sang his praises at the opening ceremony.

God. So strange, so confusing! If there was a God, where in all of blessed creation was he now? Why would God bring him, and the world, so much misery? How could a loving God allow that? After the millions he gave! But he knew. The money was nothing. He did it for Mary. It wasn't an offering to God.

He thought of a radio show from decades years before, called "Imus in the Morning." He listened to it on his morning drive. Sometimes, Imus pretended to be "Billy Sol Hargis." He still remembered the tune:

"I don't care if it rains or freezes
Long as I got my plastic Jesus
Riding on the dashboard of my car."

Reverend Hargis ran a discount house of worship in Del Rio, Texas. "Today, and today only, my friends, you can make a $50 donation for only $19 and 95 cents. Yes, friends, you pay only $19.95 and get credit with the Almighty for the full $50."

John had to smile. Those were the days, when he and Mary were young. When it was easier to laugh, and boy, had they howled when Imus pretended to be Reverend Hargis. Those were the days when they didn't have the money. The days before things got important. There had been joy in those days.

The Lamborghini swung toward the massive gates guarding his property. Facial recognition software scanned him, and other cameras confirmed his license plate. John waited for the gates to open. Nothing happened. His fancy house, the house built to shelter him and protect him from the outside world, wouldn't let him in.

John yelled. "LET ME IN!" He painfully and slowly got out of the car. He staggered to the control panel and punched in the code. Nothing happened. He cursed under his breath.

I can't take this, he thought. *Can this day get worse?*

He grabbed his phone and called Mary. "Mary, the gates won't open. Let me in."

Mary cooed, "Oh John, so glad you're home, sweetheart. What's the password?"

"I don't want to play that game. Not today. Let me in."

"Can we please have Matthew come for Christmas? Pretty please? I miss my baby. He tells me he's changed."

"I told you I'd think about it."

"Then now is a good time for you to think about it," Mary quipped teasingly. "Take your time. Take all the time you want."

John looked again at the massive gates, set in grandiose stone pillars, blocking his way. He couldn't see his house; the driveway curved inside the gates. John's money had opened many doors, but now, at the end of his life, the gates were closed.

John almost cried. "Please, Mary! Just let me in. And remind me to shoot Mark for his stupid 'smart house' system!"

"Maybe if you were nicer to him, you'd be in charge of the system."

"Please, Mary. You know what happened the last time we saw Matthew. He was high on drugs, cursed us both, and stole the silver. It almost crushed us. We can't go through that again."

"He's our baby, John. He almost died from an overdose. He's my baby. And he says he's changed."

John paused. He had done a thousand deals, and right now, it was obvious—comically obvious—that he was not in a strong negotiating position. But more important, much more important, he was scared to face Mary. Mary didn't know the cancer was terminal. He owed her. Anything he could do for her, he had to do. And what had happened in the city had begun to change him. He wasn't in control of the universe.

"Alright. You win. Matthew can come home for Christmas."

"Oh John, you're a sweetheart! Now, what's the password?"

"My wife is the best." "Keep going," laughed Mary.

John couldn't help but let out a chuckle of his own. "My wife is the best. She is beautiful and smart, and I'm lucky to have her."

"Go on," Mary chirped. "What's the third verse?" "And I love her."

"Love you, too," said Mary. "I'm in the kitchen." The gates opened, and the speakers announced:

"Welcome home, Mr. Arnold." Then he heard Bing Crosby sing, "I'm dreaming of a white Christmas…"

"Stupid smart house. I'm going to shoot Mark."

CHAPTER THREE

"Elvis is in the building."

Mary smiled. Their son Mark, the tech whiz, insisted their house be super "smart" and have every bit of the latest home gadgets and software. Mark built his first company selling smart technology for houses and used his parents' home as a testing ground. He controlled the in-house systems and built-in quirks, such as the "Elvis is in the building" greeting. Referring to his father as "Elvis" was Mark's way of not so subtly making fun of his father's power and presence. By programming the house to announce the greeting, Mark was implying that, when his father came home, the show was just beginning. Mary waited in the kitchen. It was no ordinary kitchen. All 950 square feet screamed wealth, with its twenty-feet-high arched mahogany beams, double-thick marble, two large islands, custom oak cabinets, and museum quality art. It even had a fireplace and a sofa in one corner.

Mary didn't have to wait long before her husband arrived. She saw pain etched across John's face as he walked in.

"You're home early. Got your text. What did the doctor say?"

"Can I get a hug? Right now, I need a hug more than anything."

Mary got up and wrapped her arms around him. He seemed hot—perhaps he had a fever. He was unsteady and thinner than ever.

John looked like he was about to cry, which was astonishing. He was not a sentimental guy. "I don't feel good."

"You don't look good." Mary pressed, "What did the doctor say?"

John knew there was no way around this. This was going to hurt, and he would give anything not to hurt Mary. But he had no choice.

"Stage four cancer."

There was silence. John felt Mary slip out of their embrace, staggering back. She grabbed the kitchen island to steady herself and half fell into one of the swivel chairs. She held the counter in one hand and brought the other to her mouth in shock.

"What does that mean?" she whispered.

"I've got a tumor in my pancreas, and it's large and growing. It's spread to my liver and lungs."

Mary's eyes filled with tears. "What can they do?"
"Nothing."

Mary swallowed hard and squared her shoulders. "We'll fix this. We'll get a second opinion."

John's agony hit a new level. He had destroyed his own life, betrayed himself to arrogance, he could accept that. But in telling the truth, as he had to do, he had now destroyed and betrayed his precious Mary.

"I'm sorry, my love, but I've done that. Got a second and a third. I just met with three cancer doctors. They gave me pills for pain and pills to keep me going. There is nothing anyone can do. I don't have long. This will be my last Christmas."

Mary bent her head and let out a long wail. John was silent. At that moment, if he could have taken her pain, he would have. But he could do nothing.

"John!" Mary finally shouted. "You stubborn fool! You ornery idiot! Why didn't you tell me? And why did you wait so long to see a doctor?" He could understand her hysteria. If their positions were reversed, he would've been the same. Mary continued, tears flowing, "There must be *something* we can do?"

John managed a faint smile. "Let's open a bottle of wine." He shuffled to the wall cabinet where some of their wine was stored, and chose a bottle of a very expensive Napa cabernet. He opened it, poured two glasses, and walked to the kitchen fireplace. He pushed the switch on the side and propane flames shot up. Mary followed, and they sat side by side on the sofa in front of the fireplace and, out of pure habit, clinked their glasses together. The familiar and intimate action almost broke Mary.

"Did the doctor say you could have alcohol?" "What's it going to do, kill me?" John drained his glass.

She joined him in gulping down her drink.

As alcohol loosened their emotions, husband and wife began to talk and cry. Mary's anger was apparent, and she berated John for his stubbornness in refusing to see a doctor earlier when he might have had a chance for more time. When Iron Mary was angry, life was not good. But John was too broken to defend himself.

As the wine approached empty, he felt the need to unburden. He took Mary's hand and looked her in the eye. "Something strange happened. It happened in the city on my way home."

Mary looked confused. "What do you mean?"

"I got a glimpse of something, something beyond myself. A glimpse of something divine. A glimpse that maybe my entire life—all this struggling to prove myself to other people, to be a success and make money—isn't important. A glimpse there is something much more wonderful. I know it sounds crazy, but it was more real than anything I've ever known. It was terrifying, but then I felt peace and joy."

"You're dying and you felt joy? Seriously?" She scoffed.

"I don't know. I just felt it. It was a gift. Could this cancer be a gift? I felt love and peace. It was real. Something, or somebody, told me what I had to do, and told me that, at the end of this suffering, at the end of my life, I would be alright."

Mary looked at him. Compassion overtook anger. "Well, that proves it! You're touched in the head. Touched by an angel, you fool, is that what you think?"

"Maybe, yes."

"People will think you've gone crazy. And I'm surprised. For years, I've asked you to come to church. Only days I can count on are Christmas and Easter. Mother's Day usually. Once a month if I push."

"I don't care what people think." John's anger had evaporated. "In the little time I have left, I'm going to change. And I've decided to change my will."

"How?"

"I'm going to give the money to God. All of it. That's what I was told to do in the city, when I went to another place. And that's what I'm going to do."

"To God? What about me? The kids?"

"Don't worry." John reached over and held her shoulder fondly. "I'll make sure you're okay. Of course, I will. You never need to worry. But how much money does the brat pack need?"

Mary couldn't help but smile. It was their private joke—how much money did their spoiled, aloof kids need? The brat pack; the grubby group. They weren't sure whether leaving the kids money would help them or curse them. Money changes people, and nobody knew that better than John and Mary.

"I love you, John, even though you are grumpy, stubborn, and touched in the head. And a fool for not seeing a doctor earlier."

Mary paused. "But I don't know what's going to upset the kids more, you dying or you giving away what they think they're entitled to. This will rock their worlds. They are going to be shocked and furious. Well, the kids need to be here. All of them. I'm going to call and ask them to come. Right now. Not just Matthew, Ashley and Mark too."

John smiled. "Good luck. Christmas is two days away. But don't tell them I'm dying. Please. I want to tell them in person. I'll do that if they come. Just tell them we'd like to see them. Maybe tell them we have to see them. No, here's what you do, you tell them I *insist* they come home. Then we see who shows up. How about that? Should be interesting. Which kids will come?"

When the bottle of wine was finished, John hobbled to the bedroom, holding on to walls and furniture. His head was spinning from the wine, the drugs, and the cancer. He took a pain pill. Just one.

Mary stayed by the fire and began to call the children. She decided she was going to follow John's wishes and not tell the kids he was about to die. But she was going to do everything she could to get them to come home. Out of habit, more than any conscious order, she started with Ashley, their oldest—their visionary daughter cut from a different cloth.

CHAPTER FOUR

Ashley was in her kitchen cleaning up when she got her mother's call. She picked up on the third ring.

"Hi Mom! You and Dad ready for Christmas? Got that big house all decorated?"

Ashley was an associate professor of mathematics at MIT. That didn't mean, as she liked to joke, that she could add. Her field was theoretical math, what Einstein called "the poetry of ideas." She thought of herself as an artist and a dreamer. Her dream was to become one of the great mathematicians of history—like Archimedes, Newton, Gauss, Euler, and the mystical Ramanujan. Ashley had a mind of penetrating brightness: she could see patterns others could not see, and solve problems others could not solve. Yet, she was sweet and light on the outside.

"I think so. Ashley, I need something."

Ashley heard something in her mother's voice. This was not an ordinary call.

"What is it?"

"I need you to come home, as soon as you can, and spend Christmas with us. It's important."

"What? Mom! I told you we were hoping to visit you this spring. We're skiing in Aspen with Hajid's family."

Hajid Akbas was Ashley's husband. He was Turkish and Muslim by heritage, and a full professor of physics at Harvard, a world-renown heavyweight both in cosmology, the study of the origin and the history of the universe, and in quantum physics, the study of matter and energy at the most fundamental level of existence. The difference in religious backgrounds didn't bother Ashley and Hajid; she wasn't much of a Christian and he wasn't much of a Muslim. They thought of themselves as open- minded and enlightened humble beings. Sort of a "yes, we're ordinary people, but keep in mind, we have fancy degrees and are at fancy schools, and we can't help it if we're smarter than you, and, by the way, it is true if you need to know that we have globs of money, but we're going to leave you guessing because we're not going to tell you where it comes from." In their own eyes, they were highly moral. Of course, being smart people, they were entitled to define morality as it suited them.

While nothing was perfect, they were well matched. When they met, as they joked to their friends, it was equations at first sight. But what held them together was common values. Both appreciated the value of education—lots of it. One could never have too much education; the more degrees and awards, the better. Both respected people with academic credentials—that was an absolute given. In their view, if someone had academic credentials you could trust them because they knew what they were talking about. They accepted, with great humility, their duty to be part of the intellectual aristocracy.

Mary repeated, "I need you to come home for Christmas."

"Mom, we have tickets. We leave early tomorrow morning to escape the snowstorm. Have you been watching? It's going to be massive."

"Ashley, please come home. Get here as soon as you can, and spend Christmas with us. It's important. You'll never forgive yourself if you don't. You must be here. Trust me."

Again, there was something in her mother's voice. "I don't understand. What's up?"

"Trust me."

Ashley was confused. It didn't compute. "Let me talk to Hajid. Are my brothers going to be there?"

"I called you first; I hope they will."

"Obnoxious Mark with his tech money and bimbo girlfriend? He texted me a picture to brag about her. Looks like artificial body parts."

"Come on, Ashley. You don't know her. You've never met her. Give her a chance."

"Ugh. What about Matthew? Is he out of jail?" "Yes. He says he's straight, no more drugs. And pray we don't have another screaming scene with your father." Ashley sat on one of the stools at her kitchen island, and rotated it slowly. "Mom, even if we do all make it, it'll be the strangest Christmas ever. We haven't all been together in four years." "I know," sighed Mary.

"But Rebecca is excited about the mountains with snow and skiing." Rebecca was Ashley's daughter, nine years old and utterly adorable.

Iron Mary was just warming up. She hit Ashley with the big gun: "Your father *insists*. And don't worry about snow, we're going to have plenty."

"Why does he need to see us? Why now, why right away? We've got flight tickets, ski tickets, and restaurant reservations. We'll see you at spring break."

"Ashley, listen to me. Your father *insists*."

Ashley was not used to a world where things didn't make sense. "Let me think about it."

"Think about it all you want," said Mary. "Take your time. Take all the time you want. We'll see you here Christmas Eve. Tomorrow. No presents. Please just come home."

Ashley turned to Hajid after her mother hung up. Hajid had overheard parts of the conversation and was now sitting next to Ashley. "My father says we have to be there for Christmas. And my mother wants us to change our plans. As if our plans don't matter!"

Hajid was silent. This was not good. Ashley's mother, Mary, was wonderful but her father, John, was a grumpy sourpuss who somehow, and with little to no education, had managed to make a humongous amount of money. John was not the least bit impressed by Hajid's intellectual prowess, and Hajid tried his best not to be impressed by John's money. Hajid resented that, even as a full professor, his salary and Ashley's salary combined were woefully inadequate to support their lifestyle. He resented that the only reason he and Ashley were able to live in their large renovated townhouse on a tree-lined street in Boston's Back Bay, and have ridiculously more money than any of the other professors, was because of her father's generosity. There was no other way to put it: they were addicted to her father's money.

"What about Aspen?" Hajid finally replied, with an irritated frown across his brow. "My family is expecting us."

"My mother says my father *insists*!" "Just say no."

Ashley snapped. "Do you realize how much money my father has?! We have to go. There was something in my mother's voice. I can't take a chance. I can't risk getting cut out of the will."

Ashley looked at Hajid. "I'm so angry with Mom. It's like we're being summoned home, as if we are in debt to them because they helped us buy this house and give us money."

"Don't be angry with your mother. She's just the messenger. This comes from your father. He thinks he owns us." Hajid sighed, standing up. "Great. I guess my family doesn't count. So, it's Christmas in Connecticut with the grinch?"

"I'm sorry. I can't risk getting cut out of the will. I can't."

Hajid walked off to take their ski clothes out of the suitcases and to call his family.

CHAPTER FIVE

Welcome, Mr. Arnold. Your table is ready." Mark smiled. He was on top of the world.

The views from the Michelin-starred restaurant on the 38th floor in downtown San Francisco were the best. He was tall (6'2), young (barely 30), in great shape (worked out almost every day), good-looking, and had an MBA from Stanford. And, best of all, he was rich.

His clothes reflected his status, dressed in the best money can buy. His white linen shirt complimented his dark hair well, especially paired with the grey slacks and tan Gucci loafers. His custom navy-blue cashmere sports jacket completed the look, with a little pop of color from his red silk Christmas pocket square. Anyone who looked at him would recognize his power—he made sure of that. His girlfriend Tiffany had to be seen to be believed.

On the sexist scale of one to ten, Tiffany was an eleven. The tantalizingly short red Christmas dress showcased her magnificent curves. Her natural strawberry blond hair was layered to perfection, with soft locks cascading down her back. She was tall (5'12, she liked to joke), tanned, and athletic. Her necklace sparkled with forty large diamonds.

Heads turned as they entered. Most men couldn't help but stare. Women stared, some at Tiffany, some at Mark. While Mark looked like a model, Tiffany actually was one.

Tony Bennett's music was playing softly as they walked in. "The good life, full of fun, seems to be the ideal…"

Mark came to a sudden stop. A plain-looking man was sitting against the back wall with an equally plain woman. It was Bill, an angel investor. Bill was famous for finding and funding unicorns—companies that grew quickly from nothing to be worth over a billion dollars, companies like Stripe, Instacart, Databricks, Miro, Airtable, and Juul Labs. The Bay Area boasted over 200 unicorns, with more each year. Technology was creating sudden and overwhelming wealth—sudden wealth of a type never known in the history of the world, and Bill was riding the wave. Mark grabbed Tiffany by the arm and stopped her. "I need to talk to the guy in the red plaid sports jacket.

I want him to invest." He shot her a meaningful look, "Turn on the charm. Be nice."

Mark needed cash. He was down to $20 million: a fortune for most but thin ice for Mark. Crypto was crashing, and he had lost several times that, including most of his father's investment. He needed cash and had asked Bill to invest $75 million.

"Hi Bill, great to see you! You're looking good! Merry Christmas!"

"Merry Christmas to you, too," said Bill. He stopped eating the third course of his vegetarian tasting menu. His eyes opened wide to take in Tiffany. "Pretty shade of red. Nice necklace."

Tiffany put on her magazine-cover smile and hit him hard. "Mark spoils me. Hi, I'm Tiffany. Nice to meet you."

Tiffany was from Iowa. Corn-fed, she liked to joke, especially when people commented on her height. But at age twenty-three, Tiffany was nobody's fool. When she was fifteen her divorced mother changed her name, moved them to California, and pushed her into modeling. Tiffany didn't like it, not one bit. She lost her friends and later her innocence, and she learned what men wanted. She was beginning to think she would never find a man who cared about her and not just her looks, a man who wanted to be with the simple Iowa girl inside her. Mark had shown great promise, but now, she wasn't sure. He was smart, good-looking, nice to her, and had said all the right things. But after six months the romance was fading. She was beginning to think Mark would never really love anybody but himself. Still, there was no one else in the picture, and, even though the necklace was rented, Mark did try to spoil her. As the song goes, "breaking up is hard to do."

"I should have known I'd find a smart guy like you here tonight," praised Mark. "Best Christmas decorations in town. Well, we'll let you be, just wanted to say hello." He turned to leave, before he said, "Oh, and Bill, when you get a chance, let's talk more about the crypto deal. It's got the best technology, and with the downturn now is a great time to get in. Okay, have a great Christmas! And Bill, don't overdo it on the Diet Coke. Promise me."

"I will," laughed Bill. "And please, please give my best to your father. He helped me get started, and I love doing business with him. Anytime your father's got an idea, I'm all ears. Please tell him that and give him my best."

Mark managed to hide his strained smile and simply nodded before leaving, mood now ruined because of his father yet again.

Mark and Tiffany were shown to a prized table by the window. The hostess smiled in welcome; she was pretty and dressed elegantly in black.

"Your father's secretary insisted we give you his favorite table. Enjoy your meal!"

Mark waited until she left. He was angry. "You know, it bugs me. It really does."

"What?" asked Tiffany, confused.

"Everyone talks about my father! He's grumpy and sucks the air out of the room. What about me? I'm thirty years old and starting my second company. What about me?"

"Oh Mark," said Tiffany. "It's a beautiful restaurant. It's almost Christmas. Look around and count your blessings."

Mark took a deep breath and calmed down. *She's right*, he thought. He looked left to the Golden Gate Bridge.

He looked right at the Christmas decorations and elegant people. He looked down at the fine china and sparkling crystal. He looked across at Tiffany. She was quite the prize. What man wouldn't want to be him tonight? It was going to be alright; it *had* to be alright. He needed to get it together and put his father out of his mind. A bottle of wine might help. He opened the list to the California cabernets. Better keep it under $500. Need to conserve cash. Might want two bottles.

The first bite of truffle pasta was almost in his mouth when Mark's phone vibrated. Something made him look. Well, well, well. He hated to leave the pasta, but he got up anyway. He couldn't take the call at the table and risk looking like a jerk, especially when Bill might see. He walked to the lobby and picked up on the seventh ring.

"Hi, Mom. This is a surprise. What's up?"

"Are you enjoying the season?" Mary asked. "I hear music. Are you at a party?"

"We're at that restaurant Dad recommended. Dad's secretary got us a reservation. It's the best. I'm here with Tiffany."

"When am I going to meet her?"

"Maybe spring. I've got to stay close to take care of my crypto thing. The market is being difficult."

His mother paused, as if taking a deep breath, and said, "Mark, I need you to come home right away. Don't ask questions, just come. Get on a plane right away."

"I can't. We go by private plane tomorrow to Maui. We've got a week at a villa on the ocean with Tiffany's model and celebrity friends. It's going to be awesome."

"Tell the pilot to go in the other direction. Your father insists. And do this for me. We don't ask much of you. Do this for your father and me."

"What's going on?"

"Just come. Your father *insists*." "Can you tell me more?"

"No, but hear me. Your father *insists*."

Mark let out a scoff. "Ugh. Let me think about it." "Think about it," she said. "Take your time. Take all the time you want. We'll see you Christmas Eve. Tomorrow. No presents. Please come home."

Tiffany knew something had happened from the look on Mark's face.

"My parents want me home. My mother says my father 'insists.' What a jerk—with no warning, the guy *insists*! I'm summoned home like I'm one of his employees."

"When?" "Tomorrow."

"Tomorrow? We can't! *You* can't. Maui is going to be fantastic. Did you see the pictures of that estate on the ocean? You're going to like my girlfriends, and I want to meet their celebrity boyfriends. We've been planning this for months."

Mark couldn't look her in the eye. "It's my father. He and his money. He's got to control everything."

Tiffany smiled and tried to cheer him up. "Just say no. You're going to like my girlfriends. I promise it's going to be fun."

Mark was in serious pain. Tiffany's girlfriends were some of the hottest women on the planet. But crypto was crashing, and Mark had to have money, he just had to. He could not imagine life without money. He sat in agony, contemplating his choices. But he knew there was no choice.

"I'm sorry. I can't say no to my father. I need his backing. I need him to invest more money."

Tiffany put down her glass. She glared at Mark. "Are you kidding?"

"Sorry, I have to go."

Tiffany let out a resigned sigh. "Then go. But I'm not coming with you."

CHAPTER SIX

It was almost 11:00 pm in the South Bronx. Matthew and Imani were about to eat dinner. They had a lukewarm container of hot and sour soup, a take-out box of moo shu pork with pancakes, and a cold egg roll. It was quite the feast.

Matthew washed dishes at the Chinese restaurant two blocks over. It closed at nine on weekdays. Matthew finished up a little over an hour later, and, if there was extra food, sometimes they let him take a little home. The couple that owned the restaurant were good to him, real good. They were church friends of Imani's and gave him the job even after knowing his background. It's not easy to get a job if you have a criminal record.

"Bless this food to our use and us to thy service. In Jesus's name we pray."

Matthew looked at Imani. She could hardly keep her eyes open. She worked 12-hour shifts as an aide in the emergency room. She hadn't had time to change; there were drops of blood on her scrubs. Three patients died tonight on her shift. He was glad to have her back safe. Thank the Lord. You didn't walk around the South Bronx singing carols, at least not alone. You kept your head down, walked straight from the subway, and stayed inside. Murders were up 34 percent in one year. Police protection was fading. Just standing outside could get you killed.

The apartment was 240 square feet, with a toilet that sometimes clogged and a sad excuse for a kitchen. You walked up four stories, and the halls smelled of stale cooking and worse. Quite what, Matthew didn't like to imagine. But he was thrilled. For two years, he shared a prison cell with a paranoid schizophrenic. Then he was homeless. This was a palace.

Matthew smiled. He couldn't remember the last time he was this happy. He felt blessed and thankful. Today was his 40th day of sobriety. Forty days without demons. That was an expression for some but naked truth for Matthew. Forty-one days ago he smoked cocaine that, unknown to him, was laced with fentanyl. The other two guys died. Matthew alone survived, who knows why.

When he woke up, he saw Imani and heard her singing softly. Twice, she took a few minutes from her job to go up and check on him. After he was released, he went back to ask her out. She turned him down at first. She didn't need a white guy with a criminal record. And why her? She had a scar on her face from when a man beat her up and cut her with a knife. She was a span short and a stone heavy. She grew up without a father and had two brothers, one full and one half, killed by gangs. She survived high school by sheer force of will and, when she wasn't working 12-hour shifts, took nursing classes.

In her spare time, Imani sang in the choir. Matthew had never heard such a beautiful voice. He had wanted to be a singer; he had a band in high school. People said he had talent. But he played with drugs, thinking it was innocent fun. He became an addict and dropped out of college. His parents moved him home to "fix" him. Six times they sent him to expensive clinics; six times he relapsed. To pay for the drugs, he robbed houses in the neighborhood. He pawned items from his parents' house he hoped they wouldn't notice. The final straw was when some of the silver went missing.

Imani was Matthew's pride and joy. He adored her. Just last week, she had agreed to move in with him. They were sitting on folding chairs salvaged from a dumpster. The small table they found upside down on a sidewalk. They had a cheap mattress on the floor. They had no other furniture. But they were together, and they were happy.

Imani's phone rang. Matthew had used it to call his mother, and he had given his mother the number. Imani saw who was calling and handed the phone to Matthew.

"Hi Mom. Merry Christmas."

"Merry Christmas to you. Are you okay?" "Yeah, I'm good, real good."

"Are you sober?"

"Yes, Mom. I'm great. Today is my 40th day without alcohol or drugs. Imani and I are celebrating with Chinese food."

"Is that the aide from the hospital?" "Yes."

"Where are you?"

"We have our own apartment! Can you believe it? Friends helped us." "Where is it?" "The South Bronx." "Is it safe?"

"We're fine, Mom. Fine." Matthew paused. "How's Dad?"

"Not good. Matthew, come home for Christmas. Please."

"I thought Dad never wanted to see me again. Screamed that at me, actually." "He didn't mean it." "Could have fooled me."

"Matthew, I can't wait to see you. Let's be a family. Come right away, come tomorrow. I want a family Christmas, the whole family. It would be a miracle. And I don't want any presents, except for my baby to be home and well. You're going to be my Christmas present. I've been praying for years. I can't wait. It would mean the world."

"You know, Mom, I've started praying too." "Are you saying that to make me happy?"

"It's true. Something happened. You won't believe it. Well, maybe you will, but Dad won't. Imani is the only one I've told."

"Your father might surprise you. What happened?" "I'll tell you when I see you. I'm born again. I'm changed."

"What do you mean?" "We'll talk."

"Okay. Come home for Christmas. A family Christmas."

"Will Ashley and Mark be there?" "I hope so."

"Mom, they hate me. They are embarrassed by me, and I don't blame them after what I've done."

"You're our prodigal son. Come home. Your father and I love you."

"I don't know, Mom. I'm scared. I'm more scared to come home than to walk around the South Bronx. I'm scared of the memories and that I'll be overwhelmed, that I'll want the drugs again. And I can't leave Imani. She's my rock. She helps me stay sober."

"Then bring her. Absolutely. She sounds wonderful. I want to meet her." "Mom, Imani's black."

Matthew waited, wondering how his mother would react. Mary responded immediately.

"That's wonderful. We're all children of Adam and Eve. We're the same in the eyes of God."

"Do you still have neighbors over for Christmas dinner?"

"Yes."

"That's not going to work. Imani and I won't fit in." "Don't worry. Nobody is going to give my baby attitude. All you need to do is come. I don't care what anybody thinks. This Christmas give me the best gift ever. Come home."

"Let me talk to Imani and think about it."

"No!" said Mary forcefully. "No! No thinking! Just come! Come because you love me and I love you. Come because you still love your father."

Matthew handed the phone back to Imani. "Will you come with me tomorrow to my parent's house? They want us to come for Christmas."

"Where do they live?"

"In the suburbs. Connecticut."

"Do they let people from the South Bronx up there?"

"Very funny."

"You want me to meet your parents? Ghetto girl?"

"Please. You're the strength I lean on. You're the reason I'm sober. Jesus used you to save me."

"Are your parents nice?"

"You'll see." Matthew smiled. "You be the judge of nice."

CHAPTER SEVEN

Imani was up early. There were things to do. Most important was to find a replacement for her shift on Christmas day. She called a friend from nights, who lived with three kids and her mother on the next block. The woman thanked her for calling and said she didn't mind working Christmas and needed the extra money to pay a dental bill for her six-year-old who had been bullied.

Imani called the choir director and told him she wouldn't be there Christmas Eve or Christmas Day. That was a hard call. Imani was the star of the choir, and the choir was the pride of the church. She was supposed to sing "Amazing Grace." Last year, she had sung it solo; this year, she and Matthew were to sing it together.

Imani and Matthew were to be the highlight of the Christmas Eve service. The choir director was disappointed, but he understood.

The next thing was to get time off for Matthew. She had helped him get the job washing dishes, so now she and Matthew called the owner. They told him Matthew was going back to ask his family to forgive him. The owner not only gave Matthew time off but called back after talking to his wife and offered to let them borrow their car. It was an old brown Datsun. The radio had been stolen and the antenna snapped off a decade ago. The right front bumper sagged, and there was a bad dent on the left side. It was covered with rust, and it rattled when it drove. But Imani and Matthew thanked the couple profusely. They couldn't believe their good fortune.

Imani didn't have a driver's license and had never driven, so Matthew took the wheel. They stopped on their way to get gas and were shocked at the price.

They stopped again at a thrift store run by a church in a better neighborhood. They paid $20 to get a Christmas dress for Imani. She hadn't bought a dress in two years. Matthew bought a collared shirt and grey pants for $12, and they both changed in the thrift store dressing room.

It was almost 11:00 am when they began to drive north. Imani had never been out of the city. She stared in amazement as they drove by house after house adorned with Christmas decorations and Nativity scenes, exclaiming out loud in wonder. Some houses had expensive cars parked in the driveway, and a few looked like they had swimming pools. Imani had never seen neighborhoods like the ones that were rolling by, although it was getting harder to see the houses because of the trees. It was not yet noon when Matthew took the Round Hill exit off the Merritt Parkway in North Greenwich. Imani was now speechless. Every time she was sure she had just seen the prettiest and fanciest house in the whole world, they turned a corner and there was another house, prettier and fancier. There were houses like this on television, but television was not real. These houses were real.

A few turns and seven minutes later, they were driving up to the gate guarding John and Mary's property. Geographically, they had gone 30 miles. Socially, culturally, politically, and financially, they were in a different world. The black metal gates were 10 feet high with gold embellishments and flanked by large stone pillars. Imani couldn't see the house, just trees and a driveway curving to the left.

Imani told Matthew to stop. "It's a joke, right?" Matthew gave her an embarrassed smile. "No. This is it."

"Seriously? This is where your parents live?" "Yes."

"You *cannot* be serious." "This is it."

"You should have warned me. Why didn't you warn me?"

"I didn't know what to say." Matthew looked down. "I was afraid you wouldn't come."

"But I hate rich people! You know it. They think they're better than you. They rub their money in your face. They make you feel small."

Matthew looked up, "It's going to be fine. Trust me." "You want *me* to meet your rich parents? Wearing a shabby coat and a thrift store dress? No! I don't belong

here. This is not my world."

"It's going to be fine. My mother won't care. My mother wears department store clothes and a silver cross."

"And your father?"

"He's tough, smart, and a grump, but he's not a snob. My parents used to be poor. My father makes fun of people who think they're better than others. He hates phonies. You're real, Imani."

"You bet your white bottom I'm real."

Matthew smiled. "You can do this, Imani. We can do this."

"Lord help me," said Imani. She was scared. She clenched her teeth, her heart pounding. She wanted to scream; she wanted to run. Then, suddenly, calmness and peace began to dissolve her fear. She took a few deep breaths, and she felt better. Now it was a challenge. Now she would not, could not, back down. Now she was up for whatever; now she would stand her ground and fight, fight the forces of Hell itself if she had to. She closed her eyes, bent her head, and prayed. "Though I walk through the valley of the shadow of death, I will fear no evil, for you are with me, your rod and your staff, they comfort me." [3]

"Why the 23rd Psalm? This isn't the valley of the shadow of death. This is Connecticut."

Imani looked up. "Do you know what the 23rd Psalm is about?"

"No."

"It's David's prayer when he goes down the hill, alone and without armor, to fight Goliath."

"Wasn't David a king?"

"You are so clueless when it comes to the Bible." "But I thought David was a king."

"Not when he fought Goliath. This was 3,000 years ago, about 1,000 years before Jesus. David was an ancestor of Jesus and became the most famous king the Jews ever had. David grew up with his father Jesse and seven older brothers in Bethlehem, a village five miles south of Jerusalem. They raised lambs for sacrifice to God."

"Where did you learn this?"

"Bible study. You might try reading it. It's in the book of First Samuel. I went with my mother and brothers. We loved the Bible." Imani paused. "It's all true, and tons more interesting than your fake superhero movies."

"Was David a real person?"

"Of course! The Bible is God's word, and God cannot lie. They've dug up things proving David was a real person."

"Didn't know that." Matthew smiled. He wanted Imani to keep talking, to be distracted. "Tell me more about David."

"I'll try. But you should read the Bible. There's this war. The Jews, led by their king, a guy named Saul, are on a hill on the north side. The bad guys, the Philistines, are on a hill to the south. Between these armies is *the* valley, the Valley of Elah." Imani was getting increasingly excited as she talked about the Bible.

"*The* valley?"

"Yes. *The* valley, the valley of the shadow of death. It's fifteen miles southwest of Jerusalem. If you won't read the Bible, you should go there. You can see the 3,000-year-old fortifications of the Jewish army."

"Why were they fighting? Who are the Philistines?" "Read the Bible! Anyhow, who cares? Doesn't matter. The Philistines lived on the seashore. They were scary. They had bronze armor. That was a big deal because if the enemy swords and spears couldn't break through your armor, you probably won, and the other guy probably died. And the Philistines had this guy who was their 'champion.' Sometimes, in ancient times, for countries on the sea in that part of the world, they would settle it with a one-on-one fight, a fight to the death between the toughest guy on each side."

Imani continued. "It's one of the most incredible battles in history. And there's a lot of incredible history in the Bible, if you would try reading it. But pretend you're there—3,000 years ago. You and the other Jewish fighters are dug in on the hill to the north, looking south over the valley of the shadow of death, the valley of Elah. You look out and see another army, the Philistines, across from you dug in on the hill to the south. You've built stone walls and fortifications, and the hill is steep, so if the Philistines charge, you're going to kill a lot of them. But the problem is the Philistines have pretty much done the same thing on their hill. It's a stalemate. Then this champion guy— the champion of the Philistines—walks out. He walks out into the valley, the valley of Elah, the valley of the shadow of death, to challenge the Jews."

"Is that Goliath?"

"Of course it's Goliath! Read the Bible! Book of First Samuel, Chapter 17. How could you grow up without reading the Bible?!"

Imani saw Matthew smiling. She wasn't dumb, and she thought maybe he had deliberately changed the subject, from the absurd wealth of his family to an ancient story. Whatever. She was going to teach him the Bible if she had to ram it down his throat.

Matthew spoke. "And Goliath is big, right? He's Goliath."

"He's more than big. He's the biggest, baddest ever. He's eight feet tall, maybe nine. And he's dressed in bronze. He's got a bronze helmet, bronze armor, bronze protectors on his legs, and a bronze javelin on his back. He's got a huge bronze shield. He's a killing machine."

Imani kept going. "So, Goliath walks out, by himself, into the Valley of Elah, and taunts the Jews. He's laughing at them. The Bible has him saying 'I defy the armies of Israel.' I bet he said a lot of worse stuff. Goliath does this every morning and every evening for forty days. The Jews are freaking out. They're scared and ain't nobody going down to fight Goliath. They're feeling like wimps, and they have to listen to this monster Goliath who yells insults at them twice a day. The Bible says they were 'greatly afraid.' [4] Then David shows up."

"Why?"

"Who cares! I think it's in the Bible, but I forget. Maybe his dad told him to bring food. David shows up and hears yelling, so he goes right to the front line. David sees Goliath walk out and hurl insults, saying things like your God wears army boots. David says 'Who is this schmuck?'"

"David said 'schmuck'?"

"Maybe not that exact word. David says something like, 'who does this idiot think he is? Who is this idiot who thinks he can defy the living God?' [5] That's it, now I remember, David says, 'how can this guy defy the living God?' Yes, 'defy the living God,' that's what the Bible tells us David said. You see, David doesn't see this as an ordinary fight. To David, it's spiritual warfare. It's a direct challenge to God, the living God, and David cannot and will not let that be. So, David speaks. He says send me, I'll go, I can take him."

"They let him?"

"No way! They all laugh! The kid's not even in the regular army! He's never been in battle! He has no military training and no experience. His oldest brother is furious. Tells David he's a snotty, evil kid trying to show off. I get that. David's older brothers are fighting for their lives, scared to death, listening to this monster Goliath yelling at them, taunting them, twice a day. They're tense, they're exhausted, and they're being laughed at, humiliated, twice a day. And cute baby brother shows up and says I got this.

You'd be angry too. David is embarrassing the family. But David doesn't give up. He tells other fighters he can take Goliath."

"And they let him?'

"No way! They laugh, too."

Imani looked at Matthew. Her wonderful boyfriend, the guy she had fallen hard for, dangerously hard for, the guy she felt connected with more than any person in the world, really had no clue when it came to the Bible. That's okay, she thought, I can teach him. But how could anyone with a decent education be ignorant of the Bible? What could be more important to learn? The Bible was breathed by God. It was an operating manual for life. But it was a joy to teach him. Imani shone when talking about the Bible.

Imani continued. "Somehow, word gets to the king, Saul, that a kid wants to fight Goliath. Saul brings him in and says you cannot be serious. David says, 'Hey, I killed a lion and a bear, I can take this guy.' You wonder what Saul was thinking. Maybe he's thinking what the heck, nobody else is volunteering. Maybe some voice in his head tells him to let David try. Saul decides, okay, let's give the kid a chance. Saul puts his own armor, the king's armor, on David."

"David fights in Saul's armor?"

"No way! David says not my style. He takes Saul's armor off! David walks down the hill, into the valley of the shadow of death, with no armor, not a sword, not a knife.

He crosses a stream. That stream is still there now. It's on the north side of the valley. David picks up five smooth stones. That's because smooth stones fly straighter. David heads toward Goliath. No armor and no weapons except a slingshot and five stones."

"Sounds brave," said Matthew, impressed.

"It's beyond brave! But David is wearing the armor of God, and he knows it. Goliath laughs at this unarmed boy who has come out to fight him. But David doesn't flinch. David says, and you might try reading the Bible because the Bible's got all the details, David says, 'I'm going to cut off your head.' Then, and here's the moment, David runs at Goliath. Runs! David can't wait to kill the man who has defied the living God. David pulls out a stone and slings it into Goliath's forehead. Goliath collapses. David takes Goliath's own sword and cuts off Goliath's head."

"Wow!"

"Hey! It's *way* more than wow, it's an over-the-top, totally cool, amazing act of God! You should read the Bible."

"Anyway, the Jews see this and whoop it up! They cheer! They yell! They charge down the hill! The Philistines see this and wonder what the heck happened. They are shocked, frozen, in disbelief. A kid with no armor just killed their invincible champion, ran at him and killed him with no fear. Killed the mighty Goliath like some other kid might kill a rabbit. The Philistines decide they've had it. They're out of here. Suddenly, they're the scared ones, and they aren't sticking around for this fight. The Bible gives the names of two towns that the Jews chased the Philistines into as they slaughtered them."

"And it's all true?" Matthew smiled. This was doing a great job of distracting Imani, of taking her out of her fear.

"Of course, it's true! Like I said, it's in the Bible! Nobody could make up a fake story and put it in the Bible."

"Okay. Then what happened?"

"I really wish you would read the Bible. Saul asks for David again. David shows up with Goliath's bloody head in his hand. It's awesome! Can you imagine? The king and his generals are there, all the big shots who were too scared to fight Goliath themselves. And cute shepherd boy walks in holding the head of Goliath. It's still dripping blood! Saul wants to know more about the kid, what's his background, how he did it. David says he's the son of Jesse from Bethlehem."

"What's so important about Jesse?"

"Oh come on! Am I going to have to teach you the whole Bible? How could you not know who Jesse is? He's David's father, and he's in the Christmas story, the story of Jesus's birth. You shouldn't have wasted those years with video games and superhero movies." Imani stared at her boyfriend in disbelief, befuddled by his ignorance of the Bible. "Why didn't they teach you the word of God?!"

"I did worse things."

Imani softened. She took another deep breath. "Those days are gone. Together, with the grace of the living God, we are going to make it. And today, with the grace of the living God, wearing the armor of God, I will meet your father. Me versus Goliath. But you should have told me."

"Sorry, Imani. I love you."

"Love you more. And you should read the Bible."

Imani looked again at the massive steel gates. She was still scared but calm. She was ready. She took a slow breath, then another. "Okay," she said, "in we go. Bring it on. Ghetto girl versus Goliath."

Matthew put the car in gear and rolled up to the gates.

Imani turned to Matthew, smiled sweetly, and spoke softly. "Maybe I'll cut off his head."

CHAPTER EIGHT

*"He's making a list, checking it twice
Gonna find out who's naughty or nice …"*

T'was the day before Christmas, and all through the house….' The music was not helping John's headache. He did not have a good night. "Music off," he said loudly. An English butler's voice responded. "Your wish is my command, sir." Stupid smart house. Just what he needed, a sarcastic fake butler.

John was drinking coffee. The caffeine was kicking in, the steroids were taking effect, and the wine was fading. Matthew would be arriving any minute.

John knew to cut losses. He learned that the hard way. In the beginning, when he started his first fund, he had invested in a new company. They claimed they had technology to change the world. They lied, the lab reports were faked. But he had met and believed the founder, ignored the early rumors, and bought more—doubled down even—when the stock started to sink. Big mistake. Huge. It was a tough lesson. He learned that he couldn't fall in love with a stock; he had to be tough and know when to fold.

He had done the same with Matthew. He wanted to believe things would get better. Even after six expensive and failed trips to rehab centers, he stuck with it. But when Matthew began stealing from the neighbors' houses, even stealing silver from his own house, and then went to jail, John couldn't take it. He told himself that was it with Matthew. Lesson learned. He put on his business face and cut his losses. Like many other parents, he gave up. He cut off all contact.

But now, Matthew claimed to have changed. The old John would not have believed it possible. But now John himself was changed. He kept thinking about what had happened in the city. And the more he thought about it, the more amazed he was. He was convinced that somehow, who knows why, he had been given a glimpse of Heaven and a taste of ultimate truth.

And the truth was that giving up on Matthew was the hardest thing he had ever done. He did it to survive and to help Mary survive. But now, he was scared. Mathew had relapsed six times. Would there be a seventh?

The virtual butler announced: "Someone is at the gate.

Unknown New York plates."

"Let them in," called Mary. She turned to John. "Let's go. We need to be out front when he pulls up."

John put on his bathrobe and walked out. It was cold and grey. He had to smile. The world's ugliest car drove around the curve and came to a stop. Matthew got out. Their eyes met. John saw clarity and love; no demons. His heart swelled. The old Matthew had come home.

John lost it. He hobbled over to Matthew and embraced him tight. "Welcome home, son. Welcome home. You were lost and now are found. Welcome home. I've missed you. More than you know." He couldn't stop crying. He was shaking. He hugged Matthew tightly, for a long time.

Matthew was surprised. He had not expected this. "Dad, I messed up. I'm sorry. I'm sorry for what I did to you and Mom. I'm different now. I want to tell you about it. If you can, please forgive me."

John saw a short black woman get out of the passenger side. She was wearing an old coat with a stain on the side and an expression of fear mixed with grit. Matthew saw where John's gaze had wandered to. He pulled back from the embrace, taking hold of his father's hand and facing both his parents. "Mom and Dad, I'd like you to meet Imani. She's my rock."

Mary spoke first, loud and welcoming. "We are delighted to meet you, Imani. Delighted! You are most welcome! Thanks for coming!" She walked over and hugged a frozen Imani.

John saw the look on Imani's face. He'd seen it hundreds of times. Prejudice comes in all shapes, sizes, and colors. He was used to people assuming that because he was rich, he was a snob, bigoted, arrogant, or all of that and more. But he wasn't. He detested those qualities. He was demanding and often brisk, that was for sure. He could be one heck of a grump, and, if you crossed him in a business deal, you were probably headed to bankruptcy court. But he wasn't a snob. He knew what it was to be poor. He took people as they were, and he despised those who thought money made them better than anyone else. Fancy degrees didn't impress him; he had a lifetime of firing fancy degrees who couldn't perform. He liked people who were real, true to themselves, free of pretensions—people with integrity.

John walked around the car to Imani. "Can I give you a hug, too?" He saw Imani begin to relax and lower her armor. She nodded. He gave her a light hug. "Let's get your stuff in the house. That storm should hit soon. It's going to be a monster."

"Nice to meet you," said Imani.

Matthew opened the trunk and pulled out an ancient suitcase. Imani grabbed her purse and in they all went.

They walked through massive double doors adorned with Christmas greenery, into the large foyer with a twenty-five-foot-high ceiling, and back to the kitchen. They settled around the second island, in high swivel chairs. Mary served tuna fish sandwiches, with coleslaw, deli pickles, and barbeque potato chips. It was one of John's favorite lunches.

John felt blessed. It wasn't the food; it was the clearness in Matthew's eyes. To be sitting across from his youngest son was a gift from God. Was it related to what happened in the city? John didn't know. But he could sense there was something Matthew wanted to say.

Matthew steeled himself, and took a deep breath. "I've been to Hell."

"It must have been horrible, pure hell. I never thought I would see this day, the day when you came home. I can only imagine the hell you've been through."

Matthew shook his head. "No. Please hear me. I mean it literally. I have been to Hell. It's a real place. As real as this world." He could see that his parents were confused. "When I overdosed, I died and went to Hell. I'm not joking, and I'm not trying to be poetic. Hell is a real place." He looked his father in the eye. "I've been there."

John could see Matthew was afraid he would seem crazy, afraid he would be laughed at. John said nothing, but nodded slowly, to encourage Matthew to continue.

"I've been reading. I'm not the only person who's been to Hell and back. There's science on this. Studies. I went to Hell. I was in Hell, and demons were eating my flesh. It was as real as anything that has ever happened to me. It was not a dream or hallucination. It was real."

No one spoke. Matthew gathered courage and kept going. "They were hideous, like hellish aliens with naked, grey bodies and faces without features. The stench was overwhelming. They were eating my flesh and torturing me. Then, suddenly, Jesus was there. He banished the demons with a word and told me not to worry, that I would be okay. He told me I had to return to spread his Word. I didn't want to go; I wanted to stay with Jesus. It was the greatest high I have ever known. But Jesus told me I had to go back to save others, and that he would show me how to do that. He told me that when I returned to this world, the first person I would see would be my soulmate, the person who would help me fight my demons. And then I was back in this world, in the hospital room. I opened my eyes and saw Imani."

John could see Matthew was struggling to try to make them understand.

"Dad, I was there." He turned to Mary, "Mom, I was there. If you search the internet, there are dozens of studies, in medical journals and other places. They're called near-death experiences. Some think millions have had them. Most are pleasant and inspiring. But some are hellish. Even saints have had hellish near-death experiences. People who have had hellish experiences usually don't tell others."

Imani spoke up. "All the aides and nurses see people who talk about things like this. Some say they saw themselves floating above their bodies. They even describe things outside the room that they could not possibly have seen. But a few wake up in terror."

The silence was deafening. Matthew's face was begging for acceptance. John knew Matthew was scared they would think he was crazy, that they would dismiss him, perhaps mock him. But his father was changed, too. For the first time in years, John was proud of his son.

His son had shown the courage to come home and face his wrath. His son had risked alienating them, further disappointing his parents, and appearing crazy. John knew Matthew didn't have to do that. He had feared Matthew would act as if the past had never happened, shirk all responsibility, and ask for money. But Matthew was trembling. John knew, he absolutely knew, that this story, this seemingly impossible journey to Hell and back, had actually happened. John was overjoyed. Now he knew. The old Matthew was back, changed, but the son he had loved had returned, bared his soul, and asked for forgiveness. This *was* a gift from God.

John responded, "I believe you, son." Mary went over and hugged Matthew, eyes filled with tears. Matthew reached over and took Imani's hand.

"That took courage," stated John. "And I know it's true." John paused and smiled. "It has to be true because you're not that good an actor."

Matthew laughed through his tears. Mary grabbed a box of tissues and passed it around. For a moment, life was good.

Then John was sad. "Something strange, something wonderful, happened to me too. But first, I have to tell you something not so wonderful, something you need to know." John looked down. There was a long pause. Now, it was John's turn to gather his courage. He didn't mince his words.

"I have cancer, and I'm going to die. I'm going to die soon, real soon. Maybe in a few weeks, maybe in a few days. There's nothing the doctors can do. I won't be with you much longer."

John looked up and saw shock. There was no way for Matthew to know this was coming. Matthew bowed his head and shifted his weight. After what was less than a minute, but seemed like an eternity, Matthew looked up.

"Dad, I didn't know."

John smiled. "Of course, you didn't know, my darling and prodigal son, of course you didn't. I've been beyond stupid. At first, I wouldn't see a doctor, and then I didn't tell your mother it was serious. I thought money made me invincible. I thought I could find doctors, buy doctors if I had to, who would fix it. But the early diagnosis turned out to be correct. I have stage four pancreatic cancer. It's deadly and spreading fast. My money cannot beat death. "Yesterday I summoned the best cancer doctors in the world, to meet with me, to fix me. I flew in the best guy from Johns Hopkins and a world-famous woman from the Mayo Clinic. We spent an hour and a half in the office of the top cancer guy at Mount Sinai."

"You should have told me," said Mary sadly. "I would have been there."

John looked at Mary with the fondness of a shared lifetime. "I'm sorry, my love. The old me couldn't admit I was beaten. I didn't want to worry you. I was wrong, and there's no excuse."

"How could you arrange that?" said Imani. "At my hospital, no one gets more than a few minutes with a doctor."

John smiled. "My dear girl, I'm beginning to see why my son says you're special. You'd be surprised what private jets and a hundred million dollars in hospital endowments can do.

"I argued with the doctors for over an hour. Yelled at them, offered them billions. Tried to buy my way out of it. They kept telling me there is nothing anyone can do. They told me the cancer could not be cured, would not be cured, and would kill me in a few weeks—maybe sooner, maybe in days. They gave me pills for pain and steroids to keep me running, for a little while.

"But don't feel sorry for me. I was as low as I'd ever been, sitting in the back of the car that picked me up. For a moment, a shameful moment, I considered ending it, taking the coward's way out, taking every pain pill in the bottle. I couldn't face your mother. I couldn't believe I had been that stupid. The pain was bad.

"I told the driver to pull over. I saw no way out, I wanted to take the pills. But something made me look up. And then, it happened. Something wonderful and beyond understanding. I, too, was in another world. Then I was back. And as we were driving home, a great peace came over me. I was going to be alright. And in an instant, I knew, I knew as sure as anything I have ever known, that being rich didn't matter. Money doesn't matter. You sure can't take it with you, and money wasn't going to save me. I knew, I just knew in a way I can't describe, that I had to change my life, even if I had only a few days. To make sense of it. And I decided right then, at that moment, to give it all away. To give away a lifetime of stuff and be free of the demon of money. And that's what I'm going to do."

John saw Matthew struggling to make sense of his words.

"Dad, that's wonderful. Not the part about you dying. But the part about you being at peace is wonderful. You haven't always been the most relaxed father. But I love you, Dad, and I know you tried to love me."

Now John's eyes started to tear up. "I did. I remember I took you on bike trips when you were younger. Remember that bike trip to Newfoundland and those silly French islands on the south side? We had some good trips together."

"I remember," said Matthew. "And when we were going to take the ferry, I couldn't find my passport. My bag was too messy. I braced myself for your screaming at me, but you didn't. You were calm."

John knew then his son had really come home, free of the demon of drugs, and offered silent thanks to God.

"Hey Dad, don't worry about the money. I don't need it. I've got Imani and Jesus. I'm going to be fine. But have you told Ashley and Mark?"

"No," sighed John. "They are not going to like it. They are really not going to like it. It will be a Christmas present from Hell."

CHAPTER NINE

The snow hit an hour later. The weather reports were accurate. In less than twenty minutes, playful flakes morphed into a blizzard. Right around that time, the house received a guest. The facial recognition and license plate readers had a match, "Princess Ashley has arrived." Ashley drove her two-month-old Tesla SUV up to her usual spot, the fifth door in the eight-car garage on the side. The "smart" house had the garage door open. She pulled in, and she, Hajid, and Rebecca headed for the kitchen.

As they came around the corner, the first thing Ashley saw was a short black woman sitting at the second island. An instant later, the kitchen burst with joyful reunion.

Mary ran up and gave Rebecca a huge hug, and there were gasps of delight and smiles all around. Her mother looked great, thought Ashley, but her father was thin and pale. Maybe he had a bug. But, and this was also out of place, *really* out of place, he had a big smile on his face, a relaxed and genuine smile that reached all the way to his ears. Then she turned to Matthew. He looked relaxed and happy, too. What the heck? There was something about him, a shine to his eyes, as though her real brother might be back—the lovable punk she used to have so much fun with before the drugs. But where did he get such bad clothes? And what about the black woman? Were they a couple? So many questions in an instant, so many unknowns in the equations she needed to solve.

In the middle of her thoughts, the house announced, "A black limousine is at the gate. Unable to identify."

"Let them in," called Mary. "Open garage door eight." Ashley was confused. *Who shows up in a snowstorm?*

Mark? Not likely. What force could move that mountain of California ego to Connecticut?

As if reading her thoughts, Mary turned to her. "That could be Mark. I'll check. You all stay here. It's going to be a special Christmas."

It's going to be a weird Christmas, thought Ashley. She looked at Matthew. "Are you well? What's going on?" Okay, it was a lame question, but she didn't know what to say.

"I'm great," smiled Matthew. "I've been blessed. I'd like you to meet Imani. She brought me to Jesus."

The equations were changing. New unknowns. Blessed? Jesus? Who was this poorly dressed guy who looked like her kid brother? Why was he with a short overweight black woman with a scar on her face? Did the drugs cause permanent damage?

Her mother came back accompanied by Mark and a woman. Immediately, on sight, Ashley hated the woman.

Pure instant subconscious prejudice. No woman should be that good-looking. It wasn't fair. Ashley liked to think she was good-looking, and she really was, but compared to this woman, she was chopped liver, maybe worse. In an instant, Ashley took it all in, the designer clothes, the necklace with large diamonds, the long strawberry blonde hair, the magnificent curves, and, worst of all, the face.

She thought she recognized that face from magazines. Deep down, the insecure little girl in her hated the woman who walked in with Mark.

As Mark spoke, Ashley forced a smile. *Nothing changed there*, she thought. *Same old Mark. Who cared that he and Tiffany—was that a real name or a joke—were going to Maui but switched plans and chartered a faster jet to beat the snowstorm? Who cared they were going to be with Hollywood A-listers? Was that supposed to impress her? Who cared his crypto company was expanding, and blah blah blah about that.*

Then, the gears clicked. Wasn't crypto crashing? Of course! Mark needed money; why else would he reroute his chartered jet from Hawaii to Connecticut? He was here to get money from Dad. Of course! He might even try to get it all and cut out Ashley. She was going to have to be careful. Well, she had solved one equation. Tiffany was eye candy to sweeten up Dad. Not that Dad had a roving eye, far from it, but no man could ignore a woman like that. This was serious. A fortune was at stake. There were still unknowns, but she had solved the main equation, or at least she thought she had. What to do? Well, start by being nice. Play the game.

"Delighted to meet you, Tiffany. I'm Ashley." She walked over and gave Tiffany a big smile and a peck on the cheek. "Welcome to our humble family." That line always brought a smile. Humble? Not really.

Tiffany smiled. It seemed genuine. "Thanks, Ashley. I've heard wonderful things about you."

Ashley figured that was a lie; she couldn't remember the last time Mark said anything remotely nice about her. Didn't he call her the dizzy princess, the royal you-know-what in the behind? And that was when he was being nice. Ashley watched as Tiffany made the rounds, worked the room, saying hello and stunning them with her smile and conversation. Barbie seemed to have a brain; was that possible? And this black woman—what was her name, Imani?—fit in like a pile of dirt in a jeweler's window. Not that she was being the least bit judgmental or prejudicial, of course. You had to be smart. You had to take the world as it is. You had to protect yourself so you didn't get cheated out of your fair share. A lot was at stake, a fortune. The wheels in her mind were spinning, analyzing, trying to compute.

It got worse. Both Imani and Tiffany were talking to Rebecca, and everyone was laughing and smiling. She was going to have to be very careful there. Rebecca was going to grow up with class and be a professor like Ashley, not some trashy model type. Rebecca needed to know what was important, what to emulate, and it sure wasn't looking like a bimbo or talking about Jesus. Brains and education - that was the ticket. And wait, look at Hajid! That look in his eyes as he spoke to Tiffany? *Watch out girl,* she told herself. She was going to wise him up when they were alone. He could find himself on the street peddling cosmology books. She hadn't won prizes and awards for nothing. She could see through this farce of a happy family. She was smarter than all of them.

CHAPTER TEN

There is beauty beyond understanding in a snow-storm. Beauty in the unique symmetric pattern of every flake, beauty in the falling snow, beauty in the soft white blanket covering the land.

They were gathered in the family room. It was next to the kitchen, with a twenty-five-foot opening between the two immense rooms. The scene outside the eighteen-foot-high bay windows seemed like something out of a fairytale. Gusts of swirling snow danced on a stage of shimmering white that covered the lawn, the shrubs, the trees, and the perfectly placed now-frozen pond. The interior woodwork was glowing. The ceiling had been torn out of a castle in Scotland. To the right of the window was a wall of rough granite blocks with a massive propane fireplace. In the corner next to the fireplace, but opposite the windows, was the Christmas tree. At eight feet, it looked small in the large room. Mary wouldn't have a larger tree, said it would be pretentious. The tree was mostly home-decorated, with ornaments made when the kids were small and some with special meanings. Christmas greenery was everywhere; the tree was artificial but the greenery was real and filled the house with the fresh Christmas smell of fir. The house was playing John Rutter seasonal music, and all seemed right with the world.

John sat in his leather recliner, smiling in content, for music was the spirit of Christmas, and Rutter was the star on top of the tree. Not garbage like that some of that radio stuff. "Last Christmas"—that sure was a painful song. Why did they keep playing it? The world was beautiful beyond understanding. But this was his last Christmas. He needed to tell them.

Mary had made more tuna fish sandwiches, and everyone had eaten. The bags had been brought in and taken upstairs to the guest bedrooms—the nicer guest bedrooms on the north side of the house, the bedrooms that used to belong to the kids when they were small. When Matthew went to jail, John ripped out everything kid-like and remodeled down to the studs and then some, because the memories triggered by their bedrooms were too painful. The remodeled rooms were spectacular, but the childhood furniture and pictures were gone. Of course, each child remembered well where their room used to be. Because Matthew was the first to arrive, Mary gave him and Imani the prize bedroom overlooking the pond, where Ashley's bedroom had once been. Mary had whispered to him about Ashley not seeming pleased, and John and Mary had shared a smile. Kids. Would they ever grow up? Beauty and joy were all around. Just look out the window.

Maybe it was the snow. Maybe it was the drugs. Maybe it was his family being together again, with new faces to delight in. For whatever reason, John was at peace. Not a superficial, I'm-getting-by-for-now peace but a deep sense of letting go and acceptance. Yes, he was going to die. Yes, it would be soon. Yes, this world would pass. It was okay. He had pain, but he could handle it, at least for the moment. The peace beyond understanding, the peace that came to him driving out of the city, was upon him again. He knew what he had to do.

John put up his hand and signaled for attention. "I thank all of you for coming. It is a gift to your mother and I, the perfect Christmas gift." He looked at Mary for strength. How did she do it, get everyone here in twenty-four hours? Wonder woman, that was his Mary.

"There are things I need to tell you." John stared at the floor. There was a moment of awkward silence. "I know I don't look good. You've all expressed concern for my health." John paused.

"You're right." He licked his lips nervously. This was hard, but he had to say it. "I'm dying." He paused. More silence.

John took a deep breath and forced himself to lift his eyes to the questioning stares of his family. "I've been an idiot; I've been a horrible, stupid, sad, stubborn idiot of a fool. I knew I was sick, and I wouldn't, I *couldn't*, admit it. I spent extra time in the office to distract myself and pretend things were fine. I did more deals and made extra money, as if I didn't have enough. I wouldn't see a doctor. Maybe I was scared. For sure I was stubborn, for sure I was arrogant. But there are things you can't run from, and things money or work can't change. I have stage four pancreatic cancer. I met yesterday with three top cancer specialists. They said, every one of them, as forcefully as they could, that it is hopeless, terminal, and there is nothing they could do. They gave me drugs for pain and drugs to keep me going. I don't know how much time I have, maybe a week or two, maybe days. The cancer is spreading fast."

No one spoke.

"Yesterday, I came home early, after the doctors. For the first time in my life, I was suicidal. I was in agony, and there was no way out. The pain was brutal, and I had been indescribably stupid. I told the driver to pull over. He found a space on Amsterdam Avenue, near Columbia University.

"I thought of ending it right there, with a massive overdose of pain pills. I sat in the back of my armored car, protected from the outside world but defenseless against the agony within. And it just so happened, who knows why, that we were parked in front of a church, the Cathedral of Saint John the Divine, the largest church in North America. A huge but unfinished monument to God. I was as low as I've ever been. Then something happened. Something that's never happened before."

Now John smiled. "I don't know why. Perhaps part of me prayed for mercy. But something grabbed me, and I went to a different place, a different world. There was joy and music glorious beyond belief. There were colors I'd never seen, colors more beautiful than I can describe. It was terrifying yet magnificent. I don't know where I was, or where I went to, but I was no longer in the car. Then, there was a presence, a being, beside me. Maybe it was Jesus. I don't know. It was glowing with glory. It told me everything was going to be alright, that I would soon be in a better place, and that it, this presence, would take care of me.

"I can't describe it. But something within me, perhaps my soul, exploded with gratitude toward the presence."

John paused. No one stirred, except Mary, who walked over and put her hand on his shoulder. John reached back and clasped her hand, then continued.

"The presence told me I had to go back to this world but not to worry. It told me I needed to say goodbye to those who love me. And it told me I had to break the chains that have tied me down, to break the chains of money and prestige and power and all that comes with it, to give it all away. It told me I had to rise above and push away the demons of money and power.

"So I've decided. I'm going to free myself and break the chains. This morning, while the markets were still open, I began to sell. I'm going to give it all away, hand it to God as an apology for my life. It may take weeks to sell it all without depressing the market, but it will all be sold, except for the private equity, the closely held stock, and partnership interests, which I will give away as they are. I haven't decided yet how to give it away, but I will do that very soon. I'm going to give it away, give it all away, gladly, and with great joy."

John saw Matthew smile, and stares on the faces of Ashley and Mark.

"I've already given Matthew a small amount of money to start him on a new life. Of course, your mother will never have to worry. Ashley and Mark, you are fine, you've got careers and I'm proud of you. I wish I could have been a more relaxed father for all of you. As my final gift, I'm going to set you free from the curse of heavy money. Don't worship that false god, don't bow down to that demon. The world may look at me as a success, but I charged the wrong hill. And don't think I don't love you, because I do. But this is something I have to do, something I'm going to do. I'm going to give it all, and I mean all, to God."

John stopped. Matthew wore a big smile, proud of his decision. But Mark had turned white, his brows furrowed together in a stern frown. Ashley's face showed no emotion; she stared blankly back at him. Same with Hajid. Mary squeezed his shoulder and whispered she loved him. Imani and Tiffany were smiling faintly, and Rebecca was looking at her cell phone.

John let out a small sigh of relief, having delivered the news.

"For years, I've carried around a small copy of the New Testament. I never read it, but I guess having it made me feel good, as though I was going to get blessed by osmosis, just by having it near. Yesterday, in the car, something made me take it out. I opened it and saw these words. I'd like to read them to you. It's from the fourth Chapter of Paul's second letter to the Corinthians:

"Though our outer self is wasting away, our inner self is being renewed day by day. For this light momentary affliction is preparing for us an eternal weight of glory beyond all comparison, as we look not to the things that are seen but to the things that are unseen. For the things that are seen are transient, but the things that are unseen are eternal." [6]

"Is it possible?" John wondered. "Is it possible the things we see are transient and the things we don't see are eternal?"

John looked at Ashley. He knew what she was thinking. Of all his children, her mind was most like his, the most calculating. He knew what she was thinking because he knew what he might have been thinking if he was in her shoes. Ashley's mind was churning. Game on, she was thinking. This isn't over. This is only the beginning.

CHAPTER ELEVEN

Mark was disgusted. Disgusted and enraged. Maybe more disgusted than enraged, but still enraged. Or was he more enraged than disgusted? Oh, what did it matter, who cared? His senile father had embarrassed him.

That farce about some sort of presence had ended quickly. Ten minutes later, his doped-up father almost fainted, and his mother and his sister helped a very sick man to bed and gave him more drugs. So much for the almighty Jesus. What was that refrain, something about a plastic Jesus on the dashboard of a car? Mark was open-minded—that was a given. If it made people happy to think they had a connection with Jesus, or some spirit, some sort of imaginary friend, who cared? People want to go to church? Fine, for them. But nobody should act like a babbling fool and talk about a sit down with Jesus. Be rational, please. Hadn't modern science destroyed the concept of God? Had we come so far as a species, evolved so far, yet can't give up the old superstitions? His father wasn't right in the head. That was obvious. Wait, that was it, his father was not in his right mind. Cancer and drugs had twisted him and damaged his thinking. Mark needed a lawyer. He didn't know what financial or legal steps his father had taken or planned, but Mark needed a lawyer—a smart one, and soon. He could, and he would, stop this waste of a great fortune.

Okay, so it might benefit him. Was that so bad? He was destined to be rich. He was going to be richer than his father. He had an MBA from Stanford, while his father had never gone to college. He was smarter than his father, and not that it mattered, but in all honesty, he was much better looking. He was destined to be rich, very rich, of that he was certain. He could make it happen with a good lawyer. And wait, didn't his father mention something about still thinking about how to give it away? Maybe he didn't need a lawyer, maybe he just needed to work on his father. Okay, two plans, that's the ticket. Talking his father out of it would be easier. Less messy. Ashley, the ice princess, as he used to mockingly call her, could help. And he needed help. This was important. Yes, rationality and logic, that was the ticket. Ice princess could sell that, for sure, she was nothing if not a human calculator. What a great idea! Better make that Plan A.

He texted Ashley, and twenty minutes later they met in the wine cellar in the basement. Lined with oak, the cellar held 8,000 temperature-controlled bottles. Nobody would bother them there.

"How's it going, sis?" asked Mark with a sardonic smile.

Ashley gave an equally fake smile right back. "All good, bro. All good. This is… some Christmas, wouldn't you agree?"

That was a good sign. Mark liked that; sounded like Ashley didn't fall for "the presence" nonsense either. "Dad's not well. He's on his deathbed. He's delusional. He doesn't know what he's doing."

"I couldn't agree more," exclaimed Ashley. Mark grinned. He liked the way this conversation was going. He decided to put it out there.

"I think, together, we can help him. We owe him that. We can't let him erase a lifetime of success. His legacy can't be that he lost his mind in the back of a car and wasted his money. We have to help him."

Mark could almost see the gears turning in Ashley's head. He continued, "We can't let people think our father died a fool. We can't let people think he abandoned his family. We have to protect his legacy."

"Where do you think the money should go?"

Now that was progress! Mark was almost beaming. "Well, I suppose some of it should go to science. Start by endowing fancy-name professorships at MIT and Harvard for you and Hajid. How about hundreds of millions for Hajid to search for, what is that called, 'dark matter'? He could get a Nobel Prize, right? I can see you two at the award ceremony, wouldn't that be wonderful! And you always wanted to have a chain of luxury hotels, what were you going to call them, 'royalty hotels,' or something like that? Ha! And maybe a yacht sitting in Boston harbor; Rebecca could have birthday parties on it. We can do this, Ashley!" He fist-pumped the air. "Together, we can protect Dad's legacy!"

He looked at Ashley. He knew he had overdone it, his greed was obvious, his phony speech a sham, but it didn't matter. "All of that and more is chump change," scoffed Ashley. "I want half of all the billions. We get what we can and we split it fifty-fifty."

"What about Matthew?" questioned Mark. "I don't think it's in his best interest to have money. He's fragile.

He'll be tempted to hit the drugs again. You know Imani is after his money. We have to protect him. We can do that. A few million will do that well. He'll never have to work. We can protect our little brother."

"And Mom?"

"Mom is good. She doesn't spend money. She wears department store clothes."

"And what exactly would you do with billions? Find a girl with blonder hair and bigger breasts?" Ashley's contempt was obvious.

"Ouch!" laughed Mark. "Hey, she's nice. You don't know her."

"And how's the crypto going?" Ashley raised an eyebrow at her brother, knowing she had caught him red-handed.

Mark flinched. That hurt. Ashley had done her homework. Yes, he needed the money badly, another downward bounce or two of the crypto ball and he could be bankrupt. He couldn't prop it up using the smart house company; customers were complaining about glitches in the software and the company was in the red. He squared his shoulders. Time to wrap up this deal. Appease the ice princess.

"You got it, sis. Fifty-fifty it is. We share everything equally. We're equal partners going forward. We take care of Mom and Matthew. We protect Dad's legacy."

"Yes," nodded Ashley, playing along with her brother. "It's all about protecting Dad's legacy, isn't it?"

"Look. Do you seriously want the money going to build a chain of empty cathedrals? How about a billion signs that say 'Jesus Saves'? Be rational. Let's live in this world, right now. You know this God stuff is nonsense."

"Actually," Ashley remarked, a thoughtful look on her face, "I'm not sure it is. Some people, some very smart people, think recent discoveries point to God. Of course, I'm not one of them. I believe in science and the power of human intelligence to figure everything out, but those are tough subjects. Now, don't get me wrong. I'm not saying I'm religious, and I'm not saying I'm *not* religious. I'm saying that, when it comes to God, there's room for wonder. Plenty of room. But I don't think Dad is in his right mind, and I don't think he'll put the money to good use. With my share, I can do good. I can make a difference. You spend yours as you wish. Perhaps Matthew should get more, but it could be bad for him, and destroy him.

So, fifty-fifty it is, you and I split equally what we claim as rightfully ours, and we create our own legacies. How does that sound?"

Mark's smile was so broad it seemed to split his face in two. "You got it, sis. Anything for you. We can do this. I have ideas." Mark smiled. This conversation was going well.

"Speak."

"We help Dad understand it was a delusion. It wasn't real; it was cancer and the drugs. We explain science and show the Bible is obsolete. Darwin's got the only explanation for how life evolved and the origin of species, right? We get Hajid to explain the Big Bang theory, the creation of the universe without God; isn't he the expert on how the universe was created? And we point out, and this is so obvious, that a loving God shouldn't, wouldn't, couldn't allow so much misery. When you see suffering, and we all suffer, you know God can't be real. Science and suffering, that's how we show him. Oh, and we point out that Adam and Eve weren't real people, that Jesus was a myth or a hoax, doesn't matter which, and that this resurrection stuff is nonsense. We steer the conversation to those topics."

Ashley frowned. "Some of those topics have been argued for thousands of years. On Darwin and the Big Bang, the jury is definitely out, and there are new problems. You may not win. And when Dad makes up his mind, that's usually it. He got rich doing it his way. What if Dad isn't convinced?"

Mark was ready. "Plan B. A good lawyer. Dad said something about a 'presence.' We all heard it, right? That proves he's not right in the head. If he changes his will, we get a court to annul it."

"Can we do that?"

"Not sure, but we can at least afford a good lawyer."

CHAPTER TWELVE

Mother Mary was making the best of it. She was shaken, shattered, by John's bombshell of terminal cancer. But she was delighted, over the moon, that her baby Matthew seemed to have broken the chains of addiction. And as for Ashley and Mark, well, she was always glad to see them. How wonderful they came for Christmas. Perhaps they could learn to appreciate each other, maybe even be nice to each other, and they sure better be nice in front of John if they knew what was good for them. John was not going to spend his last moments on Earth listening to fighting and backstabbing from the brat pack.

Mary focused on food. Setting her mind to the practical tasks of planning what to feed everyone, preparing and cooking each meal, was how she was managing to cope and hang in there, one hour, one day at a time. She was a great cook, and she knew how to set a table.

Earlier that morning, before the kids arrived, she called the woman who shopped for her. Now the fridge and the pantry were full of the best food money could buy, of all manner of delicious and tempting delicacies:

seafood, meats, cheeses, fruits, vegetables, and ingredients for scrumptious desserts. She was set for Christmas Eve dinner with the family and for Christmas itself. She didn't know who would be there for Christmas. She always invited a few friends and neighbors over, that was her custom. She tried to make Christmas special for John. But with the blizzard, and the snow coming down hard, who knew if anyone would make it? She called the priest at her church and begged him to come to Christmas dinner. John needed prayer and a blessing. But again, would the priest, or anyone else, be able to make it through the blizzard? Perhaps she had ordered too much food. But then she thought, who cares? She hated any form of waste, especially food. It was written in her soul from when she was poor. But then she thought, right now, with everything that was going on and her world crashing, who cares about wasted food?

She was going to focus on the soothing rituals of cooking. She had to, or she would break down. For tonight, Christmas Eve, she was making lobster pot pie. She loved the scent of tarragon—like licorice mint and celery all rolled into one. She loved the crispness of the pastry shell, the shine of the rich cream, the fresh vegetables, and of course, best of all, the sweetness of the lobster. And plenty of it. Nobody ever said she skimped on lobster. Oysters were chilling. Champagne and white wine were on ice. Brussel sprouts with bacon, one of John's favorites, was cooking on the stove. And for dessert, there was strawberry shortcake, another favorite, with real whipped cream, of course. She was going to put out a charcuterie board with the best meats and cheeses. She was going to focus on making the table look beautiful and the food delicious and try to enjoy this precious time with her kids and her granddaughter.

Just then, Rebecca walked in. So precious! Nine years old and utterly adorable. Not that Mary was the least bi-ased. Perhaps Rebecca was a little spoiled, but if so, that was Ashley's fault. And if Rebecca was a bit precocious, well that was to be expected, that was from Mary's side of the family. Mary could trace her family, on her father's side, all the way back to the Mayflower. She was descended from three of the original pilgrims. They had no wealth— her father had been a mailman—but her family had pride.

"Hi Gram!" said Rebecca. "Do you have cookies?"

Grandmother Mary loved this game. They always played it when Rebecca visited. "It so happens I do! A freshly baked one with gooey chocolate chips. How does that sound? And how about vanilla ice-cream with it? You're too thin. Don't your parents feed you?"

"Awesome," said Rebecca. "Gram, you're the best. I wish I could see you every Christmas. This is the best Christmas ever."

Mary sat her up at the first island, so she would have Rebecca close to her while she was cooking. She placed a large cookie, still warm from the oven, in a bowl and put a huge scoop of vanilla ice cream on top of it, then handed

Rebecca a spoon. Her heart clenched with so much love, she almost cried. It was too much. She had done the same for her kids when they were small. Rebecca's blue eyes looked at her for an instant, then flashed down as her spoon hit the top of the ice cream.

"Where's your mom and dad, love?"

Rebecca took a while to answer, preoccupied as she was with the cookie and ice-cream. "Dad's on Zoom. I don't know where Mom is."

Mary hummed in reply and asked, "What do you want for Christmas?"

"Same as last year. A baby brother."

You and me both, thought Mary, *you and me both.* "How's chances?"

"Not good," sighed Rebecca, as if burdened with the weight of the world.

"Anything else?"

"Well…," Rebecca thought for a second, "actually, I'd like a doll. Mom tells me that's sexist. But still, I'd like a doll. I like to pretend, and I don't see anything wrong with being a girl. I don't want clothes. I mean, that's the biggest rip-off! They're going to buy me clothes anyway! Why do they wrap up clothes and pretend it's a special gift? Do they think I'm stupid?"

Mary laughed softly before changing the subject. "How's school?"

"It's okay."

"How's math?" Mary knew Ashley wanted a daughter just like her and was pushing Rebecca hard, especially in math. Mary hoped it wasn't too hard.

"Math's okay. I'm studying Euclidean geometry mostly. But non-Euclidean geometry is cool. Did you know you get different geometries, different worlds depending on your starting assumptions? It's cool. Everything depends on what you start with. You decide whether parallel lines meet or don't meet, and bang, that's all you need to do, and you get different worlds, a flat piece of paper or the surface of a ball. Anyway, Mom says I have to get real good in geometry before I can go back to algebra. I like algebra. Then she's going to let me do calculus. I don't think calculus is going to be that hard, do you?"

Mary could hardly keep from laughing. "Not for you, Rebecca. Not for you."

"Gram, what's the password for the Wi-Fi?" "Hedgefund," said Mary. "All one word and lower case." Rebecca got out her phone. Mary spelled it. "We use that password for everything. Easy to remember that way."

"Including for the house system?" "Yes."

"Cool," replied Rebecca. "That's v-e-r-y interesting."

CHAPTER THIRTEEN

Tiffany was confused. She was confused about a lot of things, but mostly about Mark. Just last night, when they were at that fancy restaurant, he seemed vulnerable. He begged her to come to Connecticut with him and meet his parents, said he loved her and needed her. But did he? He told her the necklace was hers, that yes, he had rented it to start with, but once he saw it on her, he could never return it. But was it, *should* it be, hers? Tiffany didn't know what to think. She wanted someone to love, and when Mark said he needed her and gave her a big speech telling her that she was the one, that he loved her, she changed her mind.

She came to Connecticut, even though her agent yelled at her, said she should be at the Maui house for her career. The necklace was nice, pretty over-the-top impressive actually, worth over a million dollars. Tiffany felt she had to accept. She didn't want to seem rude, or ungrateful, not after Mark's wonderful words, but she sure hoped Mark wasn't thinking he could buy her, that she came because of the necklace. She came to see if Mark was the one. She came because Mark made her feel guilty.

She hated to abandon him if he really needed her. But did he? Mark had disappeared, so Tiffany figured she could do what she wanted. Wonderful smells were wafting from downstairs. She summoned her courage and walked down the curved staircase to the kitchen, finding Mary cooking and Rebecca sitting eating.

"Hello," greeted Tiffany. "Thought I'd come down and say hi. Smells delicious."

Mary looked at her and smiled warmly. "I'm so excited you're here! I've been so wanting to meet you. I didn't get much of a chance when you first walked in. We're so glad you came. How was your trip?"

Tiffany relaxed, pleasantly surprised at the warmth of Mary's greeting. "It was good. Mark said we had to hurry so we changed jets and left early. The flight was pretty smooth, although the pilot had to go around the storm. It's great to be here and meet you." She looked out the windows at the swirling white. "And this storm is wonderful, magical! I love snow! I don't see it much anymore, but I love it. It reminds me of when I was small. And this is an incredible house to watch a snowstorm from, just perfect. Your windows are to die for."

Tiffany saw Mary wince and realized her mistake. "I'm so sorry!" Her remorse was apparent. "I wasn't thinking. Bad expression."

"It's okay," soothed Mary, though Tiffany could see her struggling. To learn your husband is dying and then have a house full of people? That had to be hard.

"Where are you from?" asked Mary, clearly trying to smooth things over.

Tiffany relaxed a little more. Mary seemed genuine, kind, approachable, more so than her own mother. "I'm from Iowa. We moved to California when I was fifteen, my mom and I. She wanted me to be a model."

"Do you like that, being a model?"

Now that was one complicated question, thought Tiffany. "I do and I don't. I like making good money and supporting myself. I like fancy clothes, and it's nice when people tell me I'm beautiful. But there's a dark side. Sometimes, actually most of the time, it all seems a bit phony, or a lot phony, and sometimes totally phony."

Mary nodded. The warmth and understanding that emanated from her was so palpable that, for the first time in years, Tiffany decided to open up.

"If I'm really honest," she confided, "most of the time I feel like I'm living in a shell. There's this outer appearance, the way I look and the clothes I wear. And then there's the inner me. In modeling, no one cares about the inner me, who I am and what I care about. They only look at my shell, the outside of me. It's not real. In Iowa, people grow corn. Corn is real." She sighed. "Most of the time, modeling doesn't seem real."

Rebecca chimed in. "My mother says you're not real." Tiffany smiled. What an adorable child. Maybe, someday, she would have a daughter like Rebecca. Some of her friends from Iowa were married now and had children.

"How am I not real?"
"My mother says your breasts are fake."

CHAPTER FOURTEEN

It was funny, but it wasn't. The conversation became awkward. Rebecca finished her cookie and went to play on her cell phone. Mary apologized to Tiffany. "It's okay," said Tiffany. "I can take it. That's the way the world works. People are jealous, people cut you down. They say I'm too thin, they say I'm too fat. They say my chest is too big, they say my hair is too long." Tiffany tried to smile, but Mary saw sadness. It was like Tiffany had been slapped inside. Tiffany let her gaze fall to the floor and spoke softly, "Your granddaughter is precious. Don't be upset with her."

Mary was moved. Tiffany wasn't an empty shell. Inside was a person. Mary's smile was genuine. "I'm not, but I will talk to my daughter."

Tiffany kept her gaze lowered, and Mary tried to comfort her. "Still, I'm sorry. I understand how you feel. I'm no model. I'm a housewife. But people are jealous of me and my husband. You may not have noticed, but we have a bit of money."

That made Tiffany look up and grin. "Really? Who knew? I guess the Monet over the kitchen fireplace should have been a clue."

That was good, thought Mary, Tiffany was moving past the sadness. She could laugh at the absurdity of life.

"Yes," said Mary, "that's a giveaway. But in the same way you think of your looks as a shell, I think of our money as a shell. It's not the real me. It has consumed my husband, he's a good man, but he's addicted to making money. I've tried to help him and be there through his tough times. But anyone who thinks money buys happiness has never had a lot of money. Because if you do, if you have a lot of money, you know money is not what it's about. You get something, you buy something, and you feel good. Then it fades. And you are left wondering. What next? You know it's empty. Yet the world, our society, everything you read, every magazine, every commercial tells you that if you have money, you have the dream life. Let me assure you, John and I do not have the dream life. In the same way being on magazine covers doesn't make your life perfect, having money doesn't make our lives perfect."

Now Mary turned sad. She looked down and took a breath, then looked back at Tiffany. "That's how I feel. In this town, a lot of people make their lives about money. They think it buys happiness. But it doesn't. My husband is dying. No amount of money can save him."

Mary saw Tiffany look at her with kindness. They were sharing the truths of their lives, the good and the bad, and it was bringing them together.

"Then tell me," said Tiffany gently, "what matters? If it's not money, or looks, or status, what is it?"

CHAPTER FIFTEEN

Mark was on cloud nine. Finally! Finally, after years of biding his time, being patient, being content with mere millions, it was going to happen. He was going to be one of the wealthiest people in the world. He would be famous. He would have it all. Billions and billions. His face on magazine covers, more famous than Tiffany! He wouldn't have to play second fiddle when people recognized her or talked about his father. He could buy a sports team, get his name out there, show people how much money he had.

What else? Mark's mind spun. Visions of wealth danced. How about a ranch, thousands of acres, in the mountains? Montana? Idaho? Utah? Mark pulled out a piece of paper and started making notes. A friend of his knew a guy who knew a guy. Mark wrote it down. "Ranch."

What else? Yes, why not? Mark wrote "Sports team" on the list. This was the best, so exciting! What else?

Of course! "Yacht!" Mark added that to the list. What kind? Powerboat? Sailboat? He knew a guy with a sailboat with carbon-fiber sails. Very fast, very sleek. Yes, a sailboat, with carbon-fiber sails, but bigger than his friend's, 200 feet long, maybe 300, and with all the latest technology, of course. Then Mark could sail it himself, control all the sails, and rule the seas, in a monster yacht of spectacular technology! People would have to notice him. Maybe race it? No, it would be too big, too one-of- a-kind, too special to race. For that, he needed a smaller sailboat, maybe 50, 60, 70-feet long, built solely for speed. Mark crossed out "Yacht" and wrote "Sailboat." He looked at it for a moment and added an "s." "Sailboats."

This was wonderful. The perfect life. He would be famous; he would be respected. It would no longer be about his father. Sure, he might miss the guy, perhaps, not likely. But even if he did, that was okay. Mark was tough, and to tell the truth, because he was a guy who liked to get to the bottom line, his father was a disagreeable dunce.

But now it would all be about Mark, not his father. And Mark would grow the fortune, be richer than his father had ever been. He would buy companies. He would be an angel investor and get in early on tech start- ups—that was an easy way to grow billions. How hard could it be? Clearly, if bozo Bill could do it, if somebody as plain and boring as Bill could make billions in tech start-ups, it would be easy for someone as smart as Mark. Mark remembered he was still waiting for a call from Bill on the crypto thing. But now, he didn't need Bill.

Mark smiled.

What was that name from Latin and history courses? Crassus, that was it, Crassus, the richest Roman in history. Mark would be the American Crassus.

And so it went. The fantasy list grew, and the visions of wealth laughed, spun, danced. But in another dimension, a dimension not of human space or of human time, a dimension of the spirit, a demon was watching. Long ago, the demon had lured Mark, dangled before him the love of money, and Mark had swallowed the lie. Like a fisherman with a catch on the line, the demon was reeling Mark in, and salivating at the capture of a soul.

CHAPTER SIXTEEN

Mary wasn't sure how to respond. What should she say? If it wasn't money, or looks, or status, what was it? Mary was hard-pressed, close to collapse. She didn't know if she could survive John's passing. But words came.

"Do you go to church?"

Tiffany forced a smile. "No. We did when I was small, back in Iowa. We lived in a small town. My father took us to the fancy church, the church the rich people went to. We had to wear our best clothes."

Mary needed to say something. "What kind of church was it?"

"I don't remember. My mother and father argued over what church to go to, and, of course, my father won. I remember what my mother said. She called it the church of the partly true and said they had the perfect minister. Herbert Waffle was his name, Reverend Waffle. My mother used to laugh."

Tiffany smiled as forgotten memories came to the surface. "My mother laughed at that name, Reverend Waffle. She said it fit, because that church and Reverend Waffle couldn't make up their mind about whether the Bible was true. We sang praises to Jesus. But you didn't talk about other parts of the Bible, like Adam and Eve or the Flood of Noah, and if you did, you didn't take it seriously. People acted like the beginning of the Bible was made up. My mother grew up in a church where everyone trusted that all of God's word was true. My mother and father argued about that. My mother used to ask him, 'At what point does the Bible start being true? If Noah wasn't real, why did Jesus talk about him?' My father couldn't answer, which made him mad. But then, he got mad over a lot of things."

Mary saw Tiffany take a deep breath and look at her. "I liked that church. I sang in the children's choir." "What happened?"

"My father left us. It wasn't over the church. He was a salesman, and a young woman two counties over liked what he was selling. Maybe he just wanted a change. It broke my heart, it broke my mother's heart, and then, we had no money.

"My mother was forced to clean houses. People were mean to her. They looked down on her and treated her like she was nothing. The church people were the worst. After the divorce, they began to say terrible things about her. We had to buy our clothes from the thrift store, which meant all the kids at school made fun of me. My best friend wasn't allowed to invite me to her birthday parties. "My mother never recovered from the shame. We moved to California and never went to church again." "I'm sorry," said Mary, sympathetic. "What they did to you and your mother wasn't Christian. How is your mother now?"

Mary saw Tiffany try to be brave. "She passed last year. Cancer."

"I'm sorry." Mary gave Tiffany a sad smile. "It must have been hard."

Tiffany paused, then looked straight at Mary. "I read an article recently. They did a study of my generation and found most people my age have a vague sense of God. They think there probably is a God, but they think all God wants is for them to be happy. So, they chase something they think of as happiness, an image of happiness from TV shows and movies, of good-looking people spending money. It's like that old song, 'The good life, full of fun, seems to be the ideal.' They try to be 'good' people, or at least they think they should try to be 'good' people, and they think that's all God wants. [7] God is their cosmic butler, and when they pray, they ask for material things. Janis Joplin sang it. 'O Lord won't you buy me a Mercedes Benz.'"

Mary knew the words and sang the next line. "My friends all have Porsches, I must make amends."

Mary looked at Tiffany. They sang together:

"Worked hard, all my life, no help from my friends.
So O Lord, won't you buy me a Mercedes Benz."

Their voices were off-key, the harmony worse. But they laughed. It was wonderful. At that moment, it didn't matter they were different—a rich woman and a model—they were souls linked in laughter.

Mary spoke when the laughter trailed off. "That's not exactly what the Bible says. Not exactly."

"I know," said Tiffany, "but it's what people my age think."

"It's a lie," sighed Mary. "A lie of the Devil. I should know."

CHAPTER SEVENTEEN

It was after six when John woke up. He was dizzy; the drugs were having a party in his head. But it was Christmas Eve, and time on the planet, for him, was running out. He forced himself out of bed, swallowed a painkiller and more steroids. Better living through chemistry, as the old slogan went. He put on a pair of comfortable old jeans and his favorite sweatshirt and slowly shuffled, still dizzy, out of the suite of rooms surrounding the master bedroom.

An innocent bystander might have mistaken the scene as normal. Rebecca, Tiffany, Matthew, and Imani were playing Clue on the second island. Rebecca was making an accusation, Miss Scarlet in the conservatory with the lead pipe, and John sensed she would win and it would not be the first time. Mark and Hajid were by the kitchen fireplace with drinks, no doubt scotches. He thought he heard Mark say something about a sports car. He grimaced. It was Mark, he remembered, who pushed him into buying the Lamborghini and the other show-off cars. He suspected that was so Mark could drive them when he came to Connecticut.

Ashley was helping Mary carry food into the dining room. On the kitchen islands were small plates of empty oyster shells and a half-eaten platter of meats and cheeses.

The house was playing "Chestnuts roasting by an open fire…." John had never had a chestnut but he'd always liked the tune. Something about it screamed Christmas.

Everyone made a fuss upon his entrance, but Mary soon steered him into the dining room and sat him at the head of the table. The dining room was spectacular. It was large and decorated colonial style, no expense spared, with wainscoting, antique furniture, and rare art. Mary put before him a plate of oysters she had saved in the refrigerator, so they were still nicely chilled. Ashley poured him a glass of white wine, a rare White Burgundy, also nicely chilled. In one of his deals, the buyer had thrown in three cases from his private vineyard in France. John recalled that was a good deal—he liked the wine and pocketed $300 million.

Everyone joined him at the table. Mary asked him to say grace. He nodded. *This God stuff is confusing*, he thought. But he was dying, and there are no atheists in foxholes, so he did his best, clasping his hands together and bowing his head.

"Lord, we thank you for bringing us together." John paused and looked around. How strange. He was struck by how wonderfully, mysteriously, and incomprehensively odd that this group was sitting in this room on this night. He saw two women he had just met, who on the surface could hardly be more different, near- polar opposites in looks, fashion, height, background, and color. He saw two aloof kids who didn't like each other, sitting rigid with stern looks on their faces, and a former addict with a smile on his face. How strange and wonderful is life.

John bowed his head and continued. He spoke slowly. He hadn't said a prayer in a long time. "Lord Jesus, thank you for this food…. Help us love one another…. Help us to rise above our troubles…. Help us rise above our doubts…."

He smiled and looked at Mary. Not bad he thought. He still had it, the old gift of tongue. He was about to throw in a few more "rise above's" when Mary kicked him under the table.

John finished in a hurry, almost jumbling up the words, "And-bless-this-food-Amen." He looked at Mary in confusion.

"My lobster pot pie was getting cold," she laughed.

John smiled. Despite the dizziness, and the pain, it was a beautiful world. How ironic that, at the end of his life, he would begin to appreciate how blessed he was, that he could begin to push away the demons of power and money. The peace he'd experienced in the city, the peace beyond understanding, came upon him again.

His gaze fell upon Ashley. His daughter looked odd. He knew that look on her face. She was calculating— what, he didn't know. But he didn't have a good feeling about it.

CHAPTER EIGHTEEN

Ashley was thinking it over. How to reach her confused father? Mark was right, the world didn't need a billion "Jesus Saves" signs, empty cathedrals, or worse. Education, human progress, science—all- powerful human intellect, that was the ticket. With her and Hajid's connections, they could make a difference. They could change the world. True, they would only get half the fortune, but that would be enough. They could further the ancient dream of a human-only utopia, of people free from old superstitions like religion, of people creating a perfect world, worshipping only the qualities of perfect human beings, and creating perfect human happiness, and perfect human love, based purely on science, the power of human intellect. Suffering was all around, nobody could dispute that, and only science could "fix" it. Only science—human beings, not God—could maximize human happiness, and what mattered more than that? The whole purpose of life was to be happy. That was a given, right? Science, rationality, and education were the keys to open a better world.

Her father should give the money to science, not God. There is only the material world; there can't possibly be anything else. What you see is what you get. You can't see God, you can't touch God, and you can't hear God, right? What was that saying—God is an unnecessary assumption? There is only the material world, or so she thought, and science rules the material world. It was up to her. Well, her and Mark, but Mark? Her brother had always been a jerk. She wasn't about to count on Mark.

She was stunned when her father said grace. What the heck? Who knew crusty Dad would melt into a Jesus freak? To make any of this work, she needed old Dad back. She had thought about Mark's plan, or plans, and decided she liked them. God only knows how an idiot like Mark could come up with a good plan, but she liked them. They had agreed. She was going to take the lead on Plan A. Talk about science. The importance of science. How it's all about science and the power of human intellect. She would show her father, prove to her father, this "God" stuff has no scientific basis. It's science that's going to save the world, not God. Rationality, not superstition.

"It's great to be together!" she gushed, hoping everyone was convinced by her manufactured enthusiasm. "What a wonderful thing to be together, and in a monster snowstorm, can you believe it!"

There were general nods and murmurs of agreement. Ashley continued. "Hey Dad, I've got to brag about my husband. You wouldn't believe what Hajid and his colleagues are up to. They're building a scientific model of the creation of the universe! Can you imagine! They're using supercomputers to show how the universe exploded from a tiny dot, the tiniest of tiny dots, a dot so small it cannot be described, and that everything, from a grain of sand to the furthest galaxies, can all be explained by science, by the laws of physics. It's a triumph of thousands of years of human intelligence to solve the question of existence. And it's all done using science. Science can now explain everything there is, how it all came to be."

"Yes," said Hajid, going along with his wife's plan. "We've done it using the laws of physics, the laws that govern how the universe works."

Imani interrupted. Lord only knows why a girl like her would open her mouth to discuss physics with professors from MIT and Harvard, but something made her speak. "Where do the laws of physics come from?"

"That's the wonder of our universe," explained Hajid. "They just are, and they cannot be violated."

Imani was surprised to hear more words coming out. It wasn't a conscious decision; she just couldn't help herself. "But how can you have laws without a lawmaker? Don't the laws of physics reveal the majesty of God? It's in the Psalms, the songs of King David. 'The Heavens reveal the glory of God.'" [8]

Imani knew her Bible, and more words gushed out. "The Hebrew word is *sapar*, which means to count or list, to make a record of. King David tells us the Heavens are an indisputable record of the majesty of God."

Ashley forced a smile. Imani was out of her league. She had no business arguing with an associate professor at MIT and a tenured professor at Harvard. But the family was watching and listening, it was Christmas Eve, and Ashley had to play nice.

"Thanks, Imani," she said, "that truly is beautiful poetry. But now, with science, we don't need the poetry. We can explain how it happened, right from the beginning, right from the very first moment, right from the Big Bang."

Tiffany's voice came as a surprise. "But where *do* the laws of physics come from? You didn't answer Imani's question. Can the laws of physics create themselves? I never thought about it, but I think Imani's right. Don't they point to a master designer, to God?"

It was disgusting. Barbie was getting under Ashley's skin. Ashley now detested the "you-know-what rhymes with witch." Barbie wasn't smart enough to see the obvious, that was a given. She shouldn't embarrass herself by pretending to have a brain. And who the heck did she and this uneducated Christian girl think they were to argue physics with her and Hajid, two of the smartest minds on the planet? But Ashley kept her fake smile on. She turned to Hajid. Her eyes said it all: help.

Hajid cleared his throat, quickly catching on. "Well. Certainly, there is confusion on that point. But it's interesting to note that Stephen Hawking, a physicist recognized by all as one of the most brilliant people of our age, and who, until his death, held the chair at Cambridge once held by Isaac Newton, didn't think so. Hawking said, 'Because there is a law like gravity, the universe can and will create itself from nothing.'"

Imani spoke again, again almost automatically, but a bit louder this time. "I've heard that!" she exclaimed. "There's a scene in a movie, that's where I've heard it. I think the movie's called 'God's Not Dead.' It's a key scene!"

She suddenly went quiet, and Matthew reached over and held her hand. Ashley could see Imani was now embarrassed to have talked out of turn and raised her voice in front of people she barely knew. *Good*, thought Ashley. But then, and Ashley could hardly believe it, it got worse. Her mother joined in.

"Tell us about it. Tell us about the movie."

All eyes turned to Imani. Ashley could see Imani did not like being the center of attention. She saw fear on Imani's face—but the fear was mixed with grit. What Ashley didn't know was that internally, Imani was putting on the armor of God. Imani took a deep breath, prayed silently, and continued.

"There's a scene. This kid goes to college and takes a course. I think philosophy."

This is ridiculous, thought Ashley. *Are you kidding me? Imani should be changing bedpans, not talking philosophy.*

Ashley was surprised Imani even knew the word! Ashley wanted to say that, to shout out and let everyone know Imani was way out of her league, expressed in a politically appropriate way, of course. But she couldn't. Her mother had asked Imani to talk. Ashley shifted her weight on her chair, looked at Imani, and smiled sweetly as if she couldn't wait to hear whatever nonsense this horribly dressed and pathetically uneducated girl was going to tell them.

Imani continued. "The professor tells every student they have to agree there is no God and sign something about that. But one kid won't. One kid won't sign, so the professor, who's a first-class you-know-what by the way, puts the kid in front of the class and hits him with that quote you just said. Yeah, something about because there's gravity, because there is a law like gravity, the universe created itself. And the kid doesn't know what to say, not right then. He's stumped." Imani paused. She was talking too much.

She's talking too much, and she knows it, thought Ashley. Maybe this is over, maybe they could get back to Hajid's model, and how science explains everything. But again, her mother spoke.

"What did he do?" asked Mary.

Ashley was shocked. This was inappropriate, seriously inappropriate. Perhaps her mother didn't realize how intellectually adept—no, let's be honest, how intellectually superior her daughter and Hajid were. They were professors at two of the world's fanciest academic institutions!

They had scores of adoring and admiring students hanging on to their every word, emulating their thoughts, and, to be honest, sucking up to them to get a good grade. And these weren't ordinary students, these were the most elite students in the country, la crème de la crème de la crème! Her mother had no clue. There were so many layers of intellectual ability, and she and Hajid had scaled the peak, made it to the mountaintop of human brainpower, and what could be more important than human intellect? Certainly not money like show-off Mark was into, or superficial looks like Barbie, or drugs like her brother Matthew was hooked on or used to be hooked on. And this girl from the South Bronx, this uneducated creature, talking about cosmology? Philosophy? Seriously? It was more than annoying; it was a violation of the natural order.

But Imani was now encouraged. Ashley could see Imani gaining strength from the knowledge that someone was interested in what she had to say.

"I don't remember exactly," admitted Imani. "Maybe the kid googles something. Anyhow, the kid finds there's this famous English mathematician who sees through the Hawking rubbish, who calls it triple nonsense, one of the stupidest things ever said by someone who people think is smart. I think the guy's name was John Lennox, not sure, a well-known professor at a big-shot English university, maybe Oxford. This mathematician asks, 'Where did gravity come from? Who put it there, if not God?'" Imani looked around the room. "That's the point, right? If God didn't create gravity, who did? And second—and this is not me talking, this is the English mathematician—how can a law of physics create something from nothing? The English mathematician says laws of physics *describe* stuff, they don't *create* stuff. And then third, finally, and this is huge, right, who lit the torch? If God didn't start it, who did?" [9]

Ashley was indignant. She wanted to reach across the table and shake some sense into Imani. What a ridiculous speech by a woman with no education! But before she had a chance to respond, to put Imani in her place, her father spoke.

"Thank you, Imani. That's a great question. You hit the nail on the head. Where *do* the laws of physics come from? Why *is* there something rather than nothing?"

Her father paused. "I used to think about things like that. Years ago, I thought about those kinds of questions. Then I got caught up making money, doing the deal, chasing the rabbit of wealth around the track. Now I'm dying. Now it matters; it matters a lot. So yes, please, let's talk about it. Where *do* the laws of physics come from? Why *is* there something rather than nothing?"

Ashley knew she had to respond, and quickly. "Some think the universe popped out of a fluctuation in the quantum field, the foundation that underlies all existence."

Her father shook his head. "Come on, Ashley. That's no answer. You know it. If it's true, where did the quantum field come from?"

Her father looked at Hajid. "You know something about this, don't you, Hajid? Is the quantum field simple?"

Ashley kept smiling on the outside, but inside, she didn't know whether to laugh or cry. Her husband was an expert in quantum physics, and her father either had no clue or was deliberately putting Hajid down, most likely the latter. It was like asking Michael Jordan if he knew something about basketball or Michelangelo if he knew something about art.

She saw Hajid try to respond calmly to the insult and try, as he had done dozens of times with her father, to gain respect.

"The quantum field is the most complex structure in all existence. At the most fundamental level, the atomic level and below, the universe melts into mathematical beauty. Fantastic mathematics—Schrodinger wave equations, Hamiltonian operators, Klein–Gordon equations, Hilbert spaces, Eigen functions, Lie algebras, and more."

"So, the foundation of the universe is complex mathematics?"

"Yes, and I am an expert on it."

"And mathematics is thoughts and ideas, right?" "You might say that."

"Then tell me," said her father, "whose thoughts are they? And why are there thoughts at all? Where did the quantum field, all this mathematical complexity and beauty, come from?"

This is backfiring, thought Ashley. Badly. Her father was looking at Hajid. They had never gotten along. To her father, Hajid was a stuffed shirt of a pompous professor, while Hajid thought her father crude and uneducated.

"I want to know, professor," said her father. "I'm not stupid and I'd appreciate a straight answer. How did the quantum field come to be? Who made it, who put it there?"

"We're not sure," gritted out Hajid. "You have to start with something that just is. Something that cannot be explained."

"Something outside the laws of physics?"

Ashley realized her father had just trapped Hajid, an intellectual checkmate. Hajid had said the laws of physics cannot be violated, and her father had made him admit there is something that is outside the laws of physics, something that does not obey the laws of physics, something physics cannot explain. Then it got worse.

"You know," hummed her father, "years ago, I saw an article. The author said there are constants of physics, numbers that have been measured, numbers that seem to be set precisely right for our universe and for life to exist. Numbers like the speed of light, and the ratio of the masses of the proton and the electron. Numbers like the strength of gravity, the strength of forces inside atoms, and so on. The author said there are dozens of these constants, perhaps hundreds, that have the exact value necessary for life to exist. Like walking into a control room for the universe and finding every dial turned to the exact right setting."

Her father gave Hajid an appraising look. "Is that true?"

Hajid thought for a moment. He decided to tell the truth.

"Actually," said Hajid, "it probably is. Stephen Hawking called it 'a remarkable fact.'"

"Isn't that evidence of God?" asked her father. "Isn't that evidence the universe was designed?"

There was silence. Hajid did not answer. Ashley had to say something.

"Dad, there's an explanation for that. The explanation is that there are an infinite number of universes with different constants of physics, different values of these constants, different values of the speed of light and the strength of gravity, and so on. We just happen to be in the one universe where the values are set perfectly right for us to exist."

"Is that what you call a 'multiverse,' this idea that there could be other universes?"

"Yes. Many scientists believe there are an infinite number of other universes, and they have different laws and constants of physics."

"And there's evidence for that?"

"Well, the evidence is that there must be or we couldn't, wouldn't, be here to talk about it."

Her father shook his head slowly and smiled at her. "Come on, Ashley! What the heck? You're talking in circles. You can't explain something in this universe, so you invent this concept that there are other universes and that there's some sort of universe-creating machine that somehow cranks them out and gives them different laws and constants of physics. Who designed that, and how did that come to be?"

Ashley grimaced, and this time, it showed. This was really, really not going well. "We don't know. It just is." "Well okay," responded her father. "I suppose that's one explanation, some sort of machine or process that magically—we don't know how—cranks out universes and magically—we don't know how—gives them different laws and constants of physics. But your idea that there are an infinite number of other universes is hard to swallow. I know enough about math to know infinity is not some cute number just out of sight on the other side of the hill; it's a monster, an alien concept that cannot exist in the real world. But okay, you can believe in your machine or whatever that somehow, and you haven't a clue how, pops out new universes."

Her father paused and his voice got softer. He looked at Mary, when he was unsure he always turned to Mary. "But isn't there another explanation? Another answer? What about God? A mind outside of space and time, a mind that created space and time, a mind that designed space and time. A mind that *controls* space and time. God. Isn't God a better answer, a better explanation for this wonder of a universe that looks in every way like it was designed, fine-turned, for life? And the God explanation, you might call it the God hypothesis, agrees with what saints and prophets have been telling us for thousands of years."

John took a moment to let that sink in. "You may think I've gone crazy; you may think I'm touched in the head."

John turned to Mary for strength, then back to Ashley. "But what happened to me yesterday, when I went to a different place and felt the love of God, that was real. It was more real than sitting with you now. I think thousands, hundreds of thousands, maybe millions of people have had something like that happen to them, something that cannot be explained. When you add it all up, all this evidence of a designed universe and everything the prophets and saints have been telling us for thousands of years, added to experiences like I had, don't you think, don't you know, that God is a better explanation? When it comes to the wonder of existence, isn't God a better answer?"

Her father turned to Hajid. "You say your colleagues have a model that explains how the universe came to be. How does the model start? What lights the spark?"

"There was a singularity," said Hajid.

"And what is a singularity?" asked her father, with perhaps a touch of sarcasm.

"Well, basically, to be honest, it's something we cannot explain."

"Something outside the laws of physics?" "Yes."

"Something that does not follow your laws that cannot be violated?"

Hajid had been checkmated a second time. Ashley could see he didn't like it. He paused and sat up straight. "Well. I suppose you could say that."

Her father looked at Mary. "Maybe it's just because of what happened to me in the city, but now that I think about it, doesn't true science point to God? What was the first cause, if not God? Where did gravity come from?"

The battle was lost. Ashley knew it. It was time to fold, time to change the subject. "Well, Dad, I guess you and I see it differently."

Ashley sighed inwardly. She had lost this round. Her father was right, when you looked at it closely, when you really thought about it, the evidence from physics pointed to God. But she had never had a relationship with God, and she couldn't, she just couldn't, get her mind around God. Most of her colleagues thought God was a stupid concept, an unnecessary assumption. Everyone who was "smart" knew that, right? They couldn't all be wrong, could they?

For Ashley, the main thing was to help her father with his money. She didn't want it to be wasted on religious nonsense. She and Hajid could do a lot of good with it; they could change the world. She decided to retreat and fight another day. Tomorrow, she would talk about Darwin, talk about suffering, and try to reach her sick father. Talk about the glory of science, and how only science, only human intellect, can save the world. Retreat for now, but don't be obvious.

"Yes," agreed Ashley. "What a nice conversation, it's all so interesting. And thanks, Mom, for this delicious dinner. Mark, would you pass the lobster pie? It's wonderful. And Imani, where did you go to school?"

"BCI."

"What's that?"

"Bronx Correctional Institute."

***

The rest of the dinner conversation was polite. About as normal as normal could be, considering. Considering that John would die soon, that the fight for his billions had begun, and that, for the most part, his children didn't like or trust or know each other. Oh, and that the household now included a gorgeous model and an uneducated (the BCI thing was a joke—and Ashley swore revenge) hospital aide who had never been out of the city. Except for those (and a few more) teensy tiny details, it was shaping up into a typical family Christmas.

CHAPTER NINETEEN

As soon as dinner ended, Mark and Ashley, along with Hajid this time, snuck off to the wine cellar for another secret meeting.

Hajid was particularly perturbed. He almost cursed. Of course, no Harvard professor would do that, and certainly not last year's winner of the Arthur D. Clampelfeather Award for Introspective Cosmology, [10] much less the distinguished co-author of the overpriced textbook his students were forced to buy, but there were rules.

There were rules, and there were reasons for the rules. And the number one rule—as anyone with a smidgen of smarts could tell you—was that you respected the intellectual accomplishments of others. You showed respect when a Harvard professor spoke. You listened to him. You learned from him! R–E–S–P–E–C–T!

And the reason—could this be more obvious?—was that intellectually superior people with academic credentials had a duty to help others, to steer them in the right direction, to protect them from inferior ideas. *If anyone is not clear on that, they can just ask me*, thought Hajid. He didn't mind explaining. It's an intellectual hierarchy, a caste system. True intellectual elites, the best of the best of the best—like he and Ashley, if you wanted a random selection—had a duty to keep humanity on the right path, to protect it from those who, okay, let's be honest, were dumber than a dead dog. And, again being honest, because he was nothing if not brutally honest and open-minded, that pretty much described 99% of the population.

At the very top, you had your tenured professors at the world's most elite universities, places like MIT and Harvard. A notch below, you had your assistant professors and full professors at less fortunate places. Below that, you were in the riffraff: doctors, lawyers, business people, and anybody else who started on the academic ladder but wasn't smart enough to make it to the top and now had to work for a living. And below all of these were the untouchables, people with little or no academic credentials. Yes, a wonderous meritocracy, but that's the way it should be, that's the way it had to be, for science to save the world. The cream rises, the best minds make it to the top, and people need to show respect if mankind is to be saved.

You see, when a tenured Harvard professor who just happens to be last year's winner of the Arthur D. Clampelfeather Award for Introspective Cosmology says the laws of physics cannot be violated, that's it. That's now science, "settled science" as he and his colleagues called it.

Ashley looked at him in defeat. "That didn't go well." Mark's anger was apparent. "Didn't go well?" he gritted out. "Are you kidding?! You got crushed! The deal was you were supposed to take the lead on Plan A, convince Dad that science is contrary to God, that science, human intellect, has disproved God, and you don't need God to explain things. And you got crushed, destroyed, embarrassed. Gravity created itself? A magical, mythological universe-creating machine that fine-tunes the laws and constants of physics? That's the best you've got? And without one scrap of evidence, not one piece of science other than imagination, and, by the way, imagination does not fit my definition of science. That's the best you've got? It was a disaster."

There was a long silence.

Ashley finally spoke up. "We can't panic. We've still got Darwin and his theory of evolution. Darwin's got the only answer for the creation of new species. And we've got suffering. Horrible suffering, unimaginable suffering, suffering a loving God couldn't, wouldn't allow. How do you get new species, Dad? Why is there suffering, Dad? Those are the questions. Dad, and that idiot Imani, will have no answers."

Mark was still angry. "I don't get it. I really don't get it. Why did you think, how could you possibly think, you were going to get away with 'gravity created itself'?"

Hajid snapped out of his sour mood.

"Sorry. I tried. I thought I could blow smoke, and I didn't expect Imani, of all people, to catch me. Of course, 'gravity created itself' is nonsense. If I don't have a clue what it means, that says something. I think Hawking spouted garbage like that to sound mysterious, to make people think he's some sort of powerful intellect above the rest of us."

Hajid turned to Ashley. "It's incredible. 'Gravity created itself' is so bad, it's not even wrong. Hawking pulled that nonsense over and over. He wrote books with nonsense in them to try to make himself into a cult genius.

And it worked. People bought millions of copies—they didn't read them, because the books made no sense. Yet people are obsessed with the guy! I hate it! I go to parties, I tell people what I do, and everybody wants to talk about Hawking. *Did you know Hawking? What do you think of Hawking? Well, I hate*—"

Ashley cut him off gently. "Hajid, dear, we need to focus."

Hajid sighed. Ashley was right. He shouldn't get carried away. He shouldn't take it personally.

"I have to admit," he stated, "and I hate to say it, but your father's not so dumb. He's right about infinity. It can't exist in the real world.

"And who could have imagined he would know about the constants of physics?"

Mark was walking around with his back to them, peering into corners with rows of incredible wines. Hajid knew Mark was doing that to show his anger, to deliberately snub them. He seethed. Ashley was right, Mark was a jerk—a jerk with no respect for him and Ashley.

Mark suddenly turned. "That was another disaster! That can't possibly be true, can it? This business that constants, things that have been measured, have been set exactly right for life to exist? I've never heard that."

"It is," said Hajid. "But if you say that, if you even hint that the evidence points to God, even in the slightest, you get laughed at. We kick you out of the club. It's academic death. You don't get funding. Unless you work at a religious college, you can't say that."

Ashley frowned.

"Hey, don't give me that look, sweetheart," he said. "You know what gets me, Hawking helped figure that out!"

"What do you mean?" asked Mark, tearing his attention away from trying to figure out how much it was going to cost to ship his father's wine.

Hajid explained. "When Hawking was young, he impressed people with his calculation that gravity is perfect, set just right for life. He calculated that if the strength of gravity was just the smallest bit weaker, or the smallest bit stronger, the universe would not have come together, or even if there was some sort of 'Big Bang,' it would have collapsed quickly. The numbers are off the charts, out of sight, more than fantastic. The odds of getting it right by chance are worse than your odds of picking the exact right grain of sand out of all the beaches and deserts all over the world, and then doing it again! Now, I don't think it was that hard to figure out, anyone who's a top professor at an elite university could have done it, but again Hawking gets the credit. That dude always, always gets the credit!"

From the corner of his eye, Hajid saw Mark wander over to the Bordeaux section. *Unbelievable,* thought Hajid. *Billions at stake, and the guy snubs us.*

"I'm telling you," Hajid continued. "Just between us, a lot of people are saying the recent discoveries in physics point to God. Over a hundred constants of physics are perfectly set. And everyone agrees, whether they believe in the Big Bang theory or not, that the universe had a beginning."

Mark looked up. "That can't be true. I've never read that." He threw Hajid a look of unwarranted distaste.

"Don't you guys have some sort of 'multiverse' theory? Something about there's lots of universes and we just got lucky? Didn't you just talk about that?"

Mark was checking out the Malbecs. What Hajid didn't know was that Mark was wondering if he could hire a temperature-controlled truck to ship them.

"You want to know what I really think about that?" Hajid questioned. "About the multiverse?" Mark did not answer. His attention had been captured by the Italian wines.

"Do you even want to know the answer?"

Mark finally tore his gaze away. He nodded at Hajid. "It's a waste of time. Guys playing with formulas and invented nonsense. Not one scrap of evidence. There's no evidence the multiverse exists, not a scrap. And even if it does exist, as three well-known scientists have figured out, you can't go back forever, which makes the whole idea useless. They proved the multiverse can't have an infinite past. [11]

"The multiverse idea is useless. It gets you nothing. You cannot avoid the question your father asked. What was the first cause? Why is there something rather than nothing? There are no answers to these questions in the laws of physics. Somehow, and we don't know how, and we sure don't know why, the universe had a beginning."

Mark smirked. "So, you geniuses don't know what you're talking about and won't admit it? Ha! Must be something in the water in Boston. We're more evolved in California. In California, we know science and religion are opposites. Sis—this science stuff, the science part of Plan A, of talking sense into Dad, you were supposed to make that pitch."

Ashley was calm. "We need to shift gears. We need to talk about Darwin. We need to explain evolution and the amazing power of natural selection. We need to show the Bible lies when it says life was created by God. That should convince Dad. And if it doesn't, we hammer on suffering. Even Imani will fold her cards on that."

Their eyes followed Mark as he wandered over to the Pinot Noirs.

"Please Mark, focus! Tonight, or tomorrow, will decide whether billions go to science or empty cathedrals. You know I'm a very rational person and treat everyone with respect and dignity. I always do that. But this uneducated Imani is a pain. Who does she think she is to argue with professors from MIT and Harvard?"

Mark's smirk grew bigger. "She nailed you. Bronx Correctional Institute! Classic! You and your phony, 'where did you go to school?' You deserved that."

"Shut up." Ashley clenched her fists.

Hajid and Ashley caught each other's eyes. *Ugh,* thought Hajid, and he knew Ashley was thinking the same. *The crazies are running the show. But we'll help them realize the importance of science. It's for their own good. Funding for my dark matter experiment, the breakthrough that's going to win me the Nobel Prize I deserve, would be nice. A yacht for Rebecca to have birthday parties on would be nice. But I'm not doing this me, I'm doing this for science.*

And I had better get some respect!

CHAPTER TWENTY

Matthew was impressed. Imani had stood up to his sister and Hajid. That took courage. But what about David and Goliath? He had to check it out. It was quite a story, and Imani told it well. But was it true?

Matthew went to his father's office and sat in front of his computer. He knew his father hated to change passwords. Sure enough, 'hedgefund' still worked. He searched 'Elah Valley' and videos popped up. He watched two: one showed the fortifications of the Israelite army from 3,000 years ago, the same ones Imani had mentioned in the car. He found a map that showed the stream where David had picked up the five smooth stones. As Imani had said, it was on the north side of the valley, so David would have had to cross it to get to the middle of the valley, to fight Goliath.

Another quick search and he was reading the Bible text: Book of First Samuel, Chapter 17. He read David's words from verse 26: "For who is this uncircumscribed Philistine, that he should defy the armies of the living God." Imani was right: to David, it was a spiritual battle. Then he read David's words from verse 45: "You come to me with a sword and a spear and with a javelin, but I come to you in the name of the LORD of hosts, the God of the armies of Israel, whom you have defied." Wow! Yep, David had put on the armor of the living God.

He was stunned by David's confidence, his trust in God.

"This day, the Lord will deliver you into my hand, and I will strike you down and cut off your head. And I will give the dead bodies of the host of the Philistines this day to the birds of the air and to the wild beasts of the earth, that all the earth may know that there is a God in Israel, and that all this assembly may know that the Lord saves not with sword and spear. For the battle is the Lord's, and he will give you into our hand." [12]

Matthew paused. There was something in that last phrase, "For the battle is the Lord's."

Matthew read all of Chapter 17, and it was pretty much as Imani had said, although the word "schmuck" was nowhere to be found. He was relieved Imani knew her stuff. But he was puzzled as to why King Saul would send an untrained youth into battle, so he did a little more digging. He was surprised to read, at the end of Chapter 16 of First Samuel, that Saul already knew David. After a mystic prophet named Samuel anointed David as God's chosen future king of Israel, the spirit of God left Saul, and Saul was tormented by evil spirits. Maybe depression, maybe demons, maybe headaches, maybe all three, thought Matthew, I've been there. Saul's people were looking for a musician to soothe Saul and help him through the dark times, and one of Saul's guys had heard David play the lyre. Cool, thought Matthew. David was a combination of Mozart and a superhero. His kind of guy. That was amazing, but it didn't answer his question. Why would a king send an unarmed musician to fight Goliath? Was that an act of God working through Saul? Or was Saul thinking, what the heck, if I let David fight and he loses, it will just look like some kid went crazy, and nobody will believe he's our toughest fighter?

But this was 3,000 years ago, more or less. They didn't have printing presses back then. Did somebody write this down, and then others pass it around and edit and embellish it so that, after 3,000 years, it's more of a fable than an actual story? But he read about how important it was to the Hebrews to copy the Bible, copy it *exactly*. It was a command from God. [13] He read about the discovery of the Dead Sea scrolls, writings from 2,000 years ago, the time of Jesus, that confirm the current Hebrew text.

He read that the Protestant Bible had sixty-six books, written by forty different authors. *Exactly* forty.

He searched more and found images of what people imagined David looked like. Of course, Michelangelo's seventeen-foot sculpture came up. Impressive. But then he saw another picture, a work by another artist, a work he recognized.

He was smiling when he walked out of his father's office. He was in awe of the Bible, the word of the living

God. He was in love with Imani. He couldn't wait to show her what he had found.

CHAPTER TWENTY-ONE

Mark stayed behind in the wine cellar. He needed to think. He couldn't believe how stupid his sister was. And Hajid? Useless. There was a reason they were in academics and he was in business.

How stupid was that Hawking quote? Gravity created itself? That was the best they had? That was how they were going to convince his father there was no God? It was rubbish, anyone could see that. And mumbling about some imaginary multiverse? Without a scrap of evidence?

His sister and Hajid were incompetent, plain and simple. They had totally fumbled the physics ball. If Ashley wanted half the money, she needed to get it together, pull out all the stops, and convince their father that reason and science are contrary to God. Perhaps she could do that with Darwin and his theory of evolution, Mark didn't know. But if Ashley couldn't get the job done, he would have to take over. But how? He did have Plan B, the challenge in court, the good lawyer. Was there another way? Was he missing something? Mark walked around the huge wine cellar, thinking it over. Then he came to a sudden stop, slapped his forehead, and laughed. Of course! He would convince his father to give him the money! He would promise that he, Mark, would use the money for God. Mark laughed out loud. So simple! And brilliant, absolutely brilliant! Yes, give the money to him, and he, the creative genius businessman of the family, would create companies and products to save the world, and would even grow the money! Why didn't he think of that before! Oh, and here's a joy, it cuts out Ashley. Perfect! Ice princess in her ivory tower wouldn't have the first clue how to create a company. It was Plan C, his secret plan—C for C-cret.

So how? Play nice. Don't attack God. Let Ashley do that. *Let her and "I'm a tenured professor at Harvard" keep on being jerks.* Hey, hey, hey--was this brilliant or what? While he, the nice child, the competent child, the only offspring who could do anything, would help his father. Yes, he would "help" his father. He would come up with ideas to spend money on things his father would approve of. He might even carry some of them out after his father died. Meanwhile, his cash crunch would be solved. This was brilliant! Yes, let the intellectual snobs lead the charge. If he had to, he could even undercut them. Ha! That would be fun. He, Mark, would show compassion to his father and help him. Brilliant, absolutely brilliant!

Now beaming with self-satisfaction, Mark looked around the wine cellar. So much wine and so little time to drink it, at least until he got it shipped to California.

He grabbed a bottle of a rare French red and headed up to check on Tiffany.

Tiffany. What a sweet woman. He was so glad she came. If any woman could help him charm Dad, it was Tiffany. And he hadn't totally lied in the restaurant.

Maybe exaggerated. He did like her. She would be hard to replace.

Tiffany was getting ready when he walked into the bedroom. "You look great, sweetie. But please wear that cashmere turtleneck I bought you instead of that baggy sweater.

Tiffany turned around, her expression pinched. "Guess what your sister says about me."

"What?"

"She says my breasts are fake. Rebecca let that slip." "Ha! That's hysterical! She's jealous! Wow, how typical of Ashley. She always has to be the best in everything."

"You can laugh, but it hurts. It makes me feel cheap." She unfolded the turtleneck Mark wanted her to wear. "I can't wear this. It's too tight. It makes my chest look too big."

That's the whole point, thought Mark. *That's why I bought it.* "Don't worry about Ashley. I'd like you to wear it. I love to see it on you."

"No," Tiffany insisted. "No way. I want them to see me as a real person."

"Alright. That's one obscenely expensive cashmere turtleneck and you look absolutely over-the-top gorgeous in it, but alright. We need to go downstairs. My mother wants us to watch a Christmas Eve service."

Tiffany relaxed a little and smiled at him. "That could be nice. Are you looking forward to it?"

Oh yes, thought Mark, *oh yes. You betcha! I will force myself to adore it. Devout son dedicated to God, that's me.*

CHAPTER TWENTY-TWO

*"Oh, the weather outside is frightful,
but inside, it's so delightful…"*

The snow kept falling, and the house, using an elaborate algorithm taking into account not only past preferences but also the time of year, the weather (both current and predicted), and the people in the house, was cranking out seasonal music. John was back in the family room, in his recliner, and at peace. The incredible peace had descended upon him again, the peace beyond understanding. Or was it the drugs? Where does inner peace come from? Can you buy it? Can you drink it? How do you get it? And what about contentment?

John was going to die. He shouldn't be feeling this happy, this contented, this much at peace. But he was. He had family with him. Death was not to be feared. It was Christmas Eve, and beauty and joy and wonder were in the world. Look around. Feel the magic. "Let it snow, let it snow, let it snow."

John loved the candlelight service on Christmas Eve. Mary never had a problem getting him to that. The Christmas Eve service was always Lessons and Carols, no sermon and no communion. Just scripture and music announcing the birth of Jesus. Not that sermons and communion were bad, but how wonderful to relax on Christmas Eve. How wonderful to relax with the words and music of the Word made flesh.

Christmas Eve service at their neighborhood church had been cancelled. Nothing was moving in the suburbs. Only a few snowplows were on the roads, and they were losing. Rebecca was helping Mary get the livestream from St. Thomas' Church on Fifth Avenue on the television. John smiled. His granddaughter was super smart. No question about it. She got that from his side of the family. The apple doesn't fall far from the tree.

The livestream was soon up and running, and the service was about to begin. The eighty-five-inch TV—he had won that argument with Mary, a rare victory, but there's no such thing as a TV that's too big, right?— focused on the altar and the wall behind. John was always amazed by that wall. It was eighty-feet high and jammed with stone carvings of saints and prophets.

John was looking forward to the music. The church's boarding school had a boys' choir. He remembered that only Saint Thomas in New York and Westminster Cathedral in London had boys' choirs of boarding students.

The snow was falling. The organ was playing. The service was about to start. It was Christmas Eve, and all was right with the world.

What could go wrong?

CHAPTER TWENTY-THREE

Mother Mary was as well as she could be, considering. Everyone was in the family room, watching the candlelight service on livestream.

The first reading was from Genesis, the first book of the Bible. It recounted the fall of mankind, when the Devil tricked Adam and Eve. God told them they could eat of every tree in the Garden of Eden, except the Tree of the Knowledge of Good and Evil. Then the Devil appeared and lied. The Devil told Adam and Eve they would be like God and have knowledge of good and evil if they ate fruit from the sacred tree. They believed the Devil and wanted to be like God, so they ate the fruit. The Devil tricked them into disobeying God. As punishment for their disobedience, God kicked Adam and Eve out of the Garden of Eden. Because of the lie of the Devil, death and suffering came into the world.

"What's this got to do with Christmas?" asked Rebecca.

"It's tradition," replied Mary.

Imani looked at Mary, seeming nervous. "I can explain, if you want me to. I teach Sunday school."

Mary was surprised, and pleasantly so. She smiled and nodded. Imani was a jewel in the rough. Maybe not exactly the woman she would have expected Matthew to bring home. But she was a jewel. A woman with solid faith and solid trust in God. "Yes please," said Mary.

This time, Imani wasn't scared. This was the Bible, and Imani knew her Bible. She knew it better than anyone in the room.

"This is the reason Jesus came," began Imani. "Death and suffering came into the world because of the sin committed by Adam and Eve. It's called the fall of mankind. Jesus came to save us, so that when he returns, we will dwell with him in a new garden for eternity."

"Cool," said Rebecca.

"That's a lovely explanation," Mark added.

"And," continued Imani, "there was a line in there that a child born of a woman will bruise the head of the Devil. That child is Jesus. From the beginning, the Bible tells us Jesus will come. He will come to defeat the Devil and end his lies."

Mary could see Ashley wasn't buying it. But then Ashley smiled. "It's a wonderful fable," she said. "Intense symbolism."

"It's not a fable," insisted Imani. "It happened. It's the truth. It's the Word of God."

"How long ago?" asked Ashley, demanding. "Where was Eden?" asked Hajid, without giving Imani a chance to respond to the first question.

There wasn't time to answer. The choir began to sing "So Long, Moses." Everyone turned to watch. Mary had never heard this hymn before, and she took care to listen to the powerful lyrics:

"And Isaiah said
He'll bear no beauty or glory
Rejected, despised
A man of such sorrow
We'll cover our eyes
He'll take up our sickness
And carry our tears
For his people
He will be pierced
He'll be crushed for our evils
Our punishment feel
And by his wounds
We will be healed
We will be healed
From you, O Bethlehem
Small among Judah
A ruler will come
Ancient and strong
From you, O Bethlehem
Small among Judah
A ruler will come
Ancient and strong
Ancient and strong"

When the last notes faded, Rebecca spoke first. "Why did they sing that?"

It was evident Imani was itching to answer. Mary gestured for her to go on, giving her silent permission.

Imani grinned and turned to Rebecca. "Those words are mostly from the prophet Isaiah. He wrote them 700 years before Jesus was born. Think about it, 700 years is a long time, a really long time, and Isaiah knew Jesus would come. You see, there's a backstory, a 2,000-year- old backstory. Jesus's coming was foretold. There were over 100 predictions, written hundreds of years before he was born, with details of how he would be born, how he would live and preach, and how he would die. Hundreds of predictions, and Jesus fulfilled every one of them."

"That's nice, Imani," Ashley interjected, "but couldn't people have written that stuff into the Bible after Jesus was born?"

Matthew interrupted. "No. The Dead Sea scrolls leave no doubt. The prophecies were written hundreds of years before Jesus."

Mary's eyes widened. She was pleased and surprised. How did Matthew know about the Dead Sea scrolls? But she didn't have time to ask. The second reading was also from Isaiah:

"There shall come forth a shoot from the stump of Jesse, and a branch from his roots shall bear fruit." [14]

Imani smiled at Matthew. "That's the Jesse I told you about, David's father, who lived in Bethlehem and raised sacrificial lambs. It's all in the Bible. Isaiah predicted Jesus would be a descendant of Jesse. Jesus' earthly parents, Mary and Joseph, were *both* descended from Jesse and David. [15] David burned out as a king, but from his burned-out stump God created a king who would rule forever."

Mary was amazed. There was more to Imani than might appear.

The church bells rang solemnly. It was the beginning of the second carol, "O Come, O Come Emmanuel". It almost made Mary cry. Of all the music celebrating Jesus' birth and Christmas, this was her favorite. It told of the longing of the Jewish people for the promised Messiah, as they waited for hundreds of years, including seventy years of captivity in Babylon:

"O come, O come, Emmanuel,
and ransom captive Israel that
mourns in lonely exile here
until the Son of God appear."

"What does 'Emmanuel' mean?" asked Rebecca.

Mary knew that one. "God with us. Jesus was God walking on the earth; God with us." Mary saw Imani smile. They were fast becoming friends.

The third reading was again from Isaiah, this time the ninth chapter, using the King James Version, beginning at the second verse:

"The people that walked in darkness have seen a great light: they that dwell in the land of the shadow of death, upon them hath the light shined." [16]

Mary looked at John. Yes, she thought, they lived in Greenwich, Connecticut, with wealth and power, but it was still the land of the shadow of death.

Four verses later came words that have resonated through the centuries:

"For unto us a child is born, unto us a son is given: and the government shall be upon his shoulder: and his name shall be called Wonderful, Counsellor, The mighty God, The everlasting Father, The Prince of Peace." [17]

"How can the government be on Jesus's shoulder?" asked Rebecca, puzzled.

Imani was there, armed with answers. "When Jesus comes back, in his second coming, he will rule the Earth for 1,000 years. It's in the last book of the Bible, the book of Revelation. The Bible begins with Genesis, the book of beginnings, the history of Creation, and ends with the book of Revelation, the book of future history, the revealing of what is to come. Isaiah is talking about Jesus's second coming, when Jesus will rule, when Jesus will have the government on his shoulder."

The third carol was "Lo! He Comes with Clouds Descending." Mary was never fond of that one. The fourth reading was from Micah.

"I thought mica was a stone," Rebecca piped up.

Imani laughed. "Funny," she said, "it's pronounced the same way, but it's the name of someone who lived about the time of Isaiah."

Another prediction from 2,700 years ago, 700 years before Jesus, was read:

"But you, O Bethlehem Ephrathah,
who are too little to be among the clans of Judah,
from you shall come forth for me
one who is to be ruler in Israel,
whose coming forth is from of old,
from ancient days." [18]

The fourth carol was "O Little Town of Bethlehem." Mary scanned the room. Mark was holding Tiffany's hand, singing along, and, surprisingly, sported a big smile. Rebecca was grinning. To Rebecca, this carol meant it wouldn't be long until she got her Christmas presents.

"Where is Bethlehem?" asked Hajid.

Imani was there, in a millisecond. "On the main road five miles south of Jerusalem. Where Jesse raised sacrificial lambs."

Wow, that was quick, thought Mary. *Impressive.*

Mary saw Ashley give Hajid a look. She wondered why.

The fifth, sixth, and seventh readings were from the Gospel of Luke, the third book in the New Testament. They told of the angel Gabriel's visit to the Virgin Mary, when she was a young girl and before she was married, telling her she had been chosen to be the mother of the Son of God. They told of Jesus's birth and how Mary wrapped him in swaddling cloths and laid him in a manger because there was no place for him in the inn. They told of the shepherds in the field that night, who were much afraid when an angel appeared. Mary liked the verses about the shepherds:

"And in the same region there were shepherds out in the field, keeping watch over their flock by night. And the angel said to them, 'Fear not, for behold, I bring you good news of great joy that will be for all the people. For unto you is born this day in the city of David a Savior, who is Christ the Lord. And this will be a sign for you: you will find a baby wrapped in swaddling cloths and lying in a manger.' And suddenly there was with the angel a multitude of the heavenly host praising God and saying, 'Glory to God in the highest, and on earth peace among those with whom he is pleased!'

When the angels went away from them into heaven, the shepherds said to one another, 'Let us go over to Bethlehem and see this thing that has happened, which the Lord has made known to us'." [19]

"This is nice," said Ashley, "but isn't it interesting that nobody knows when Jesus was born?"

Once again, Imani surprised all of them. "June 17, 2 BC."

When she felt everyone's shocked gaze on her, she was startled. "I can explain," said Imani, "I went to a talk."

Imani seemed worried she had said too much, so Mary reassured her. "I'd love to hear more, maybe later."

The eighth lesson was from the Book of Matthew, twelve verses about the Star of Bethlehem and the visit of the wise men.

"Was the Star of Bethlehem a real star?" inquired Rebecca.

This time no one was surprised when Imani answered. "I think, basically, it was the planet Jupiter," said Imani. "But there's more."

"Cool," said Rebecca. "Very cool!"

The last reading was from the Gospel of John. "In the beginning was the Word." [20] The lights in the church were dimmed, and candles were passed and lit. Voices singing "O Holy Night" filled the room. Mary was moved and delighted she had gotten all her kids to sit through a church service, even if it was on television. She wished they were all there at Saint Thomas or at least at the church down the street. Mary began to sing along. Mark quickly joined in, then everyone else. Then, for the first time, Mary heard Imani sing.

CHAPTER TWENTY-FOUR

Mark was not surprised to get a text from Ashley: *Wine cellar, five minutes.* He was deliberately late. When he finally arrived, his sister and Hajid were waiting.

The ice princess looked annoyed. Good.

"What was that?" demanded Ashley. She was standing with one hand on her hip. "What happened? You sang along? You encouraged a woman with no academic credentials? What was that? I thought we had a plan!"

Mark tried his best to appear reassuring and serious. Internally, he was having a ball. This trip home was going to be fantastic financially. Outsmarting Ashley was a bonus. A twofer!

He was ready. "Come on, sis. It's Christmas Eve. Ride the horse in the direction it's going. Act like you love it. Don't be obvious. For now, revel in this syrupy Baby Jesus, wise men, Star of Bethlehem nonsense. You know it's nonsense, but act like you love it. Focus—that's what you like to say, right?—focus on the end result. Dad's fading. You can see it. The acid of tomorrow's pain will burn through today's Sweet Jesus. Tomorrow, we convince Dad the material world is all there is, and that science, people working together without religious myths, is the only way out, the only way to relieve suffering. Tomorrow, we convince Dad to give his money to science, to let us create a lasting legacy for him and not waste the money on religion. You and Hajid will explain why science has disproved God, and why only science can ease pain and suffering. You've got the education. You've trained your whole life for this. Tomorrow will be your moment. I know you and Hajid can do it."

Mark saw Hajid wander over to the Malbecs. He hoped Hajid wasn't thinking any of that wine was going to be his.

Ashley shrugged and gave Mark a sigh. "So, we're still in this together?"

"Of course, we're together! I'm with you, Ashley.

Tomorrow. The sun will come out tomorrow."

Mark smiled inwardly. There was a reason he was the creative business genius of the family. He knew how to make things happen, get things done. Ice princess had no clue. She wouldn't know what hit her.

CHAPTER TWENTY-FIVE

They went back to the dining room for strawberry shortcake, with real whipped cream—lots of it.

Nobody ever said Mary skimped on whipped cream. John felt tired and weak, but it was strawberry shortcake, so he sat up and ate. For some things, you had to push yourself. And even more important was the family being together on Christmas Eve. It had been a beautiful service but a little overwhelming. There was so much he didn't know, so much he wanted to understand. *Wait. Did Imani mention something about Jesus's birth and the Star?* He'd be interested to find out more.

"Imani, tell us when Jesus was born and about the Star," said John. "If you don't mind."

John saw Imani look to Mary for permission. Mary nodded as a sign that Imani should go ahead.

"The short version or the long version?"

John smiled. "Maybe the medium version. The medium short version."

"Okay. I learned about it after going to a talk. Then I read books and watched videos. It makes sense. It all fits together."

Imani was feeling more at home, getting over the culture shock, and trying to get over years of prejudice against people with money. And if one thing was clear, it was that Imani loved talking about the Bible.

"People have tried to figure it out for 2,000 years. There's lots, tons of different theories. But I'll tell you what I think, if you'll let me." Imani glanced at Ashley and Hajid.

John could see Imani didn't want to be challenged by the education snobs. *Imani's got street smarts*, thought John. It was something he'd come to respect from a thousand deals. Some people, and he thought of himself as one, have street smarts, intelligence and wisdom they didn't learn at school. He expected Ashley and Hajid to challenge Imani. *Could be interesting*, he thought. *Could be fun*. Street-smart versus the education snobs. A little like David versus Goliath. Did Imani have a chance?

"The first question we need to ask," began Imani, "is when did King Herod die? This is Herod the Great."

"Why was he great?" asked Rebecca.

"Super question," complimented Imani, "but your grandfather asked for a short version, so I'll tell you all about him later."

No one else interrupted, so Imani continued. "People used to think Herod died in 4 BC," Imani went on. "But now, we know there was a copying error in an ancient book. The earliest copies of that book tell us Herod died in 1 BC." Imani paused and looked again at Ashley. "At least, that's what I think. Some people with fancy degrees don't agree."

Priceless, thought John, *priceless*.

Imani continued. "When we look at the ancient skies using computers, and we know Jesus was born before Herod died, we see that on June 17, 2 BC, and we know the exact date here, the exact date and time, Jupiter and Venus appeared to touch each other in the sky. It's called a 'planetary conjunction.' Jupiter and Venus are the brightest planets, the brightest objects in the sky after the Sun and the Moon. Jupiter and Venus touched each other, you might say kissed each other. It must have been amazing."

"Does that happen often?" asked Rebecca, with rapt attention. She loved astronomy.

"No," explained Imani. "It's only happened three times in the past 2,000 years, and June 17, 2 BC is the only time the conjunction was visible from the Middle East. It may have been the closest, the most spectacular conjunction ever. NASA supercomputers have verified this. It's true, it happened. That's part of the story of the Star of Bethlehem, what the wise men saw. But it's only part."

Ashley couldn't help herself. "But, Imani, and I don't mean to be critical in the slightest, it's very kind of you to tell us this story, but it seems a stretch to say that just because two planets appear to get near each other in the sky—and you're right that's called a conjunction, very good for you, nicely done—that just because they get near each other in the sky, that means Jesus was born that night. Planets move. Even if it happened as you say— and by the way, I agree, modern computers can show us what the ancient skies looked like at any time from any place on Earth, you did a good job on that, too—it seems a stretch to say that proves Jesus was born that night. Of course, that's if Jesus was a real person."

John saw Imani blink with shock. Imani sat up straight and shook her head in disbelief. It was as if Ashley had hit her with a bucket of cold water. Ashley had thrown down the gauntlet. The fight was on. *If* Jesus was a real person? *If? Watch out, Ashley*, thought John. *You just stepped on a hornet's nest.* John saw Imani take a deep breath, square her shoulders, and glare at Ashley. Imani was visibly angry, almost shaking.

"*First*," Imani said loudly, "I didn't say it proves anything. But now, we've got a date. Now, we can put the pieces together. For how it all fits, you need the longer version.

"*Second,* as for Jesus being a real person, perhaps you should read the Roman records. Or study history. Jesus didn't have your fancy education, and he didn't have money or political power. He was born poor, and he died poor." Imani looked around the lavish dining room. "But face facts, you can't deny it. Jesus changed the world more than any other person."

Priceless, thought John. *Perfectly priceless.* Ashley walked right into it. Imani was not going to back down. Now John needed to hear more.

"Now you've got to give us the long version," he said. "I'm staying awake for this. Hit us with your best shot. I always thought the Star of Bethlehem was a myth, made up, you know sort of like…" John looked at Rebecca and stopped. "Like a made-up story."

"You got it," saluted Imani. "Put your seat belt on." "The wise men, and let's use the word in the Bible, the

Magi, came from Saba, a city that is now called Saveh. It's sixty miles southwest of Tehran, in present-day Iran. It was a key city in the Parthian empire. The Parthians were tough. They were one of the great empires of the ancient world, an empire that lasted almost 500 years—a single dynasty. They were never conquered by Rome. They didn't practice the Jewish religion, but they had Jewish blood, and they knew Jewish history and the Old Testament. They knew the prophecies about the Messiah. Their kings claimed they were entitled to rule because they were descended from the line of David."

John was curious. "How do we know Saba's the place?"

"From Marco Polo, the Venetian trader who made it to China and back. Marco Polo saw the tombs of the Magi in Saba around the year 1300, that's almost thirteen centuries after Jesus. He gave details, said the tombs were of great size and beauty, said the bodies were well- preserved, and he even gave the names of the Magi. He wrote that the Magi left from Saba when they went to find Baby Jesus."

"Couldn't that have been added later into Marco Polo's manuscript?" asked Ashley.

John saw Imani make a face. She glared at Ashley, still angry. Imani was not going to back down. "You can always make up stuff, and come up with a theory that denies Jesus, denies the living God. You have free will. You can make stuff up. I'm trying to keep it short. Do you mind if I continue?"

John was impressed at her standing up to Ashley.

Imani gained confidence. She kept going. "Let me tell you about the Magi and the Parthian empire. The Magi were royal astronomers and the top advisors of the kingdom. If you remember Daniel, the Daniel who survived being thrown into the lion's den when God shut the mouths of the lions, that Daniel was the leader of the Magi 500 years before Jesus."

Imani turned to Matthew, who was sitting next to her. "The Magi were like Jedi knights. Very powerful, very smart, and they knew astronomy. They could predict eclipses and planetary conjunctions, and they knew Jewish history."

Again, Ashley couldn't help herself. "I don't think the Magi could possibly know in advance where planets would be."

"Then you've never heard of the Antikythera mechanism," stated Imani. "The world's oldest analog computer. Found in a shipwreck off the coast of Greece, in a ship that sank two centuries before the birth of Jesus. If the ancient Greeks could predict conjunctions, I think it's safe to say the Parthians could."

Praise the Lord, thought John. Imani was eating Ashley's lunch, besting her at every turn.

Imani looked at John. "Now the stage is set. We've got powerful royal astronomers, possibly the smartest people in the world, brilliant people with centuries of knowledge and access to the full resources of their empire. They know the Bible and they watch the skies carefully. Now we use modern computers to go back 2,000 years, and look at what they saw."

John saw Imani pause and look around the room. She was taking charge.

John saw Imani stare at Ashley. "This is going to get technical," said Imani. "Try to keep up."

Priceless, thought John, *perfectly priceless.*

"Go back nine months earlier, to September of 3 BC. The Magi witness a triple conjunction of Jupiter and the star Regulus. The Babylonians called this star 'Sharru,' meaning 'the King.' Jupiter, as everyone back then knew, was the King Planet. Jupiter passed Regulus three times and traced a crown above it. Regulus is the brightest star in the constellation of Leo the lion. The Bible compares the Israelite Tribe of Judah, the tribe of King David and Jesus, with a lion. This triple conjunction of the two kings happens only twice every 83 years."

I'm loving this, thought John. *Not sure I'm following it, but I'm loving it.*

Imani was on a roll. "On September 11, 3 BC, as the triple conjunction begins, the sun rises in the womb of the virgin in the constellation Virgo. At the foot of the sun is a slim crescent moon. Of course, the constellation wasn't visible when the sun was up. Would you agree, Professor Akbas, that you normally can't see stars in the daylight?"

Hajid huffed. "Of course! Everybody knows that." "But the Magi have eyes, the smarts, to see it. They

watch in wonder as the king moving star and the king fixed star dance together, trace a crown, in the womb of a virgin. These guys had instruments and skills we can hardly imagine." John saw Imani glance at Ashley again. "You really should read about the Antikythera mechanism. It would be a nice addition to your education."

John almost burst out laughing. Ashley shot him a dirty look, but Mark was smiling.

Imani continued. "Okay, and do try to keep up, professors, don't make me explain it again, the sun appears here only one day every year, and the moon has this shape one day each lunar cycle." She looked at Hajid. "That's 29 1/2 days, right? All signs, the Heavens themselves, point to the coming birth of a great king in Judah. This was the date of Christ's conception, this is the blessed day the Angel Gabriel appeared to Mary." [21] There was no stopping Imani. "Then, *nine months later*, it happened. God sets the prophecies in motion. On June 17, 2 BC, *nine months after* the king planet Jupiter, the king moving star, traced a crown around the king fixed star, Jupiter and

Venus unite in the sky. For the Magi, Venus symbolizes femininity, the perfect woman. The planets combine—they become one. Together, they're the brightest 'star' anyone alive had ever seen, the brightest object ever after the sun and the moon that anyone alive then, or anyone alive for the entire Parthian Empire, has ever seen. It was visible in the West, in the direction of Judah, toward Israel, for an hour after sunset.

The Savior has come. Jesus Christ is born.

"And here's a fact that blows me away, that makes me scream with adoration for the living God, a fact modern computers have confirmed. This planetary conjunction, this meeting of Venus and Jupiter, takes place again in the constellation of Virgo the Virgin. Get it? The conception occurs in the womb of the virgin, and Jesus is born from the womb of the virgin. The sacrificial 'lamb of God' is born in Bethlehem, where Jesse and David raised sacrificial lambs. Jesus is born in the early summer, not winter, during birthing season for the sacrificial lambs. The shepherds who cared for those lambs were out in the warmer weather sleeping with the flocks during birthing season. Jesus is born, and Joseph and Mary wrap him in swaddling cloths."

"What are swaddling cloths?" asked Rebecca. Imani grinned widely. "Love that question, thanks!

Swaddling cloths were old, used garments that had been worn by rabbis. They were rags, but sacred rags, probably linen. The sacrificial lambs that Jesse and David and others raised in Bethlehem were clumsy critters. To be a sacrificial lamb, they had to be perfect, without blemish. It's in the Bible; it's a command from God. [22] When a sacrificial lamb was born. the shepherds would wrap the lamb in swaddling cloths, in linen rags that had been blessed by and worn by men of faith, to keep it from hurting itself, to keep it perfect."

Imani kept going. "Imagine you're a shepherd. It's a beautiful night, and the stars are shining, and you're in the field. From nowhere, an angel pops out, and you are much afraid, you are out-of-your-mind scared. But the angel says don't worry, chill, but get yourself to Bethlehem, where you will find a baby wrapped in swaddling clothes in a manger. The angel says 'this will be a sign to you,' and it is. It tells the shepherds *exactly* where to look. They run to a place called the Tower of the Flock, a watchtower that once was part of David's royal compound, and where shepherds would bring sheep about to give birth to sacrificial lambs. You see, it's a special place. The rabbis keep it clean. It's not some filthy barn, there's no chickens or cows or goats running around. Joseph and Mary were allowed to stay in this special place because they were both descended from David. [23]

"You're a shepherd, you know exactly where to look, and you run. You cannot believe what you find. He's there, the living God. You see a baby born in the building where the difficult births of sacrificial lambs took place, a human baby born in this sacred place, wrapped in holy rags just like the shepherds wrapped a newborn baby lamb to protect it. It is a wonder, a miracle, beyond words. Two thousand years of prophecy are beginning to come true. Jesus, the sacrificial lamb of God, is sent by God to pay for our sins, an offering from God to the human race, an offering for peace between us and God. God is just, a price has to be paid for sin, for the disobedience of Adam and Eve and of every person ever born, and only God can pay it. So, God pays it himself. He sends Jesus to die on the cross for our sins. Jesus is born exactly like a sacrificial lamb, in the same town and the same place and wrapped in the same way, as a sacrificial lamb. For the important stuff, God doesn't mess around. He puts in layers and layers of symbols to help us understand. The shepherds out in the fields know this. They see a human child, wrapped in used priestly garments, in swaddling cloths, in holy rags, in a place where sacrificial lambs are born."

Imani knew she was getting repetitive but was determined to continue. "Now back to the Magi, the royal astronomers who see Jupiter and Venus touch each other on June 17, 2 BC. You can imagine. They are stunned. The Heavens have spoken. A great king has been born in Judah—a king they've been expecting for centuries, a king born from a virgin. Could this be the Messiah? They've got to check it out, they've got to see for themselves. It could be the greatest event in history, and they know it. They decide to travel 1200 miles over mountains, through bandit territory, into land controlled by Rome.

The Romans and the Parthians hate each other, and the journey is going to be dangerous. The Magi can't call a taxi, and they can't hire a private jet. They spend months preparing, knowing that to travel 1200 miles by camels and horses is going to take two months, maybe longer. They pack gold, frankincense, and myrrh to give to Jesus, really valuable stuff, gifts to give a great king. And they assemble an army. [24] Maybe hundreds, maybe thousands, of trained and armed soldiers. They put all this together, and they head out traveling west towards Jerusalem, in the direction they saw Jupiter and Venus kiss in the sky. As they travel, Jupiter points the way in the Western sky."

Imani took a deep breath and looked around. Nobody was challenging her now.

"They get to Jerusalem in late December, 2 BC. They leave the army outside the city, maybe on the other side of the Jordan River. They're not trying to conquer anyone. They ask to see Herod. They put it to him, point- blank, right in Herod's face: 'Where is he who has been born king of the Jews?'"

"Can you imagine? What a thing to say! Herod the Great is the ultimate egomaniac. He's built cities and palaces, and he rebuilt the temple in Jerusalem in spectacular glory. He was Mark Anthony's roommate in boarding school in Rome, and he's used Roman power and technology to build incredible things, things we still marvel at today. He's also an insecure, psychopathic murderer. Herod killed two of his own sons because he was afraid they might take his throne. The Magi walk up to the ultimate egomaniac, a paranoid murderer, Herod the Great, and ask to see the new king, the real king. They tell Herod they 'saw his star when it rose.'"

Imani was hitting on all cylinders, rolling full steam. "Do you get it? The Magi are showing off; they are bragging. They're dissing Herod. They're throwing it in his face! They're saying that because of their advanced knowledge of astronomy, they knew Jupiter and Venus were going to meet, and they knew exactly where to look for it rising in the morning sky. They saw this meeting of the two brightest planets 'as it rose' in broad daylight." She turned to Hajid, "Professor, would you agree this conjunction would have been visible to the naked eye, if you knew exactly where to look and blocked out the sun?"

Hajid mumbled. "Maybe."

Imani continued. "It's so cool. The Magi are bragging, showing off, showing Herod what they've got. They are totally in his face. They have skills, knowledge, star smarts far beyond anything Herod, or anyone else in the Roman Empire, had at that time. They are royal astronomers, the heirs of centuries of ancient knowledge, from a powerful empire, and they are bragging about their sophistication. Herod is clueless. He didn't even know the conjunction had occurred. He had to ask his advisors later what the Magi were talking about. Herod is clueless, and he's got some of the smartest people in the world talking to him, powerful people who show him no respect."

Imani looked at Hajid. "And that proves the Star wasn't a comet or supernova or anything like that. Herod didn't know about the conjunction. If it was a comet or a supernova, he would have known about it. But Herod had no clue.

"Herod is furious. He wants to kill the Magi, kill them right then and there, for their impertinence, for daring to ask where the real king is. He's Herod the Great, and he's just been royally dissed. And he hates the Parthians. He's fought against them. But he's got to be nice. There's an army outside the city. Herod is scared of the Parthians. He knows that fifty years earlier, the Romans invaded Parthia with seven legions, 40,000 elite soldiers of the empire that had conquered the Mediterranean world. The army was put together by the richest Roman ever, a guy named Crassus, who wanted to be the Roman version of Alexander the Great. The Parthian king sent 8,000 fighters, mostly archers on horseback, to slow down the Roman legions, to buy time while he formed a larger army. He shouldn't have worried. The outnumbered Parthians destroyed the Romans. Ten thousand Roman soldiers were killed. Ten thousand were captured, and the rest ran for their lives like the Philistines ran after David killed Goliath. Crassus was killed."

Imani paused, realizing she had gone off track. "Back to Herod. He can't start a new war. So, what does he do? He smiles, he fakes it, he plays nice. He says that's so wonderful, how truly nice of you to let me know. Thanks so much, my new friends. Please do go and find this sweet kid and tell me where he is. Yeah, says Herod, I can't wait to worship this kid too, I just can't wait, so you smart guys go and find him, and I'll be right behind you to worship him. You go ahead, I'll be right behind. Herod lies, he's faking it. Herod knows he will kill this baby the moment he is found. Ain't no snotty babe in a manger going to bring down Herod the Great.

"The Magi leave. They saw Herod was clueless, but they know where the Son of God made flesh was born. They know the words of the prophet Micah, 700 years earlier, when he foretold Jesus would be born in Bethlehem." Imani looked at Rebecca. "That's the prophet whose name sounds like a stone."

Imani composed herself as she prepared to wrap up. "The Magi head south, to Bethlehem. It's the evening of December 25, 2 BC, and they're only miles away. Remember, the Magi know the skies. Now here's something that will stop you cold. On that date, that exact date, on December 25 of the year 2 BC, Jupiter goes into retrograde motion. Jupiter stops in the sky. It stops moving against the background stars. The Star of Bethlehem stops in the sky."

Imani took a deep breath. "Finally. The Magi arrive at the house where Jesus is. Jesus is now six months old. The original Greek uses the Greek word for toddler. And the Bible tells us the Magi arrive after Jesus was born and that Jesus was staying in a house at that time. The Magi walk into a house of dirt-poor peasants. They see Jesus with Mary, his mother. Immediately, in an instant, they fall down and worship Jesus. Can you imagine? These royal advisors, the elite of the Parthian empire, the smartest people on the planet, fall down to worship the baby Jesus. They saw the Star in daylight, as it rose, they watched the planets kiss in the sky, they have come all this way, 1200 miles, with servants and a private army, to fall down and worship Jesus. They sure didn't do that when they met Herod. But Jesus is no ordinary king, and they know it, they absolutely know it. They fall down and worship Jesus, the living God, the Word made flesh. They're top guns. They represent the Parthian empire, and they're the intellectual elite of the ancient world. They throw themselves down on the dirt floor of a peasant house, and they worship Jesus. They offer gold, frankincense, and myrrh. They came 1200 miles for this moment. They fall down before the most powerful king ever, the King of Kings, the Lord of Lords, and they worship Jesus. They offer gifts. It's the first Christmas." Imani hit Ashley with a smile of victory. "And I don't care what nonsense they teach in your fancy schools.

That's how I say it went down. Any questions? Would you like me to explain the Antikythera mechanism?

Imani laughed. "Oh, I almost forgot. Two of the gifts the Magi brought were predicted by the prophet Isaiah, 700 years before Jesus was born. Isaiah foretold the Magi would come with 'a multitude of camels,' and bring gold and frankincense." [25]

John spoke up. "I thought there were three gifts?"

Imani smiled. "Yes! The Magi brought gold, a gift for a king, frankincense, which symbolizes divinity and is burnt at religious services, and myrrh, which is used in embalming and symbolizes death. Add it up, the three gifts are for a divine king who will die. The Jews did not expect a Messiah who would be put to death. God hid the final gift, the gift of myrrh, from Isaiah. The Jews did not expect their Messiah would die.

"And, in case you were wondering, 2 BC wasn't an ordinary year. It was the high-water mark of the Roman Empire. It was the 750th anniversary of the birth of Rome, more or less, and the 25th year, the Jubilee year, of Caesar Augustus as Emperor. To celebrate, the Roman Senate gave him an award and called a census."

Imani looked at Mary. "That was the census that required Mary and Joseph to return to Bethlehem, the boyhood home of King David. You see, it was the peak of the Roman Empire and, in a way, the peak of human brutality and arrogance. That was when God sent Jesus to save us."

Silence. Golden silence. "One more thing," said Imani. "The heavens are like a clock. God knew, when he flung the universe into existence, exactly when and where Jesus would be born."

Magnificent! A performance for the ages! John was blown away by the power of Imani's faith, what she knew, and how she knit it together. He pushed back his chair to stand up. His legs were weak, so he grabbed both arms to push himself up. Imani deserved a standing ovation, and he was going to start it.

He didn't make it. He lost consciousness on the way up, fell on his left side, knocked over the chair, and hit the floor hard.

CHAPTER TWENTY-SIX

Matthew woke with dawn's first light. It was Christmas. He rolled out of bed and walked to the sitting room outside the remodeled bedroom. He inserted a pod in the coffee machine: French roast with hazelnut, and he liked it black. He took the coffee mug and sat in the chair by the window, waiting for the sun to completely rise. High clouds began to shine with faint pink. Then God laid before him, slowly, brighter with each passing moment, a wonderland of incomprehensible beauty. He had made the snow fall, and he stopped it for the morning. God's white blanket glistened as the sun began to rise.

A tune came to mind, a tune his band once played:

"Morning has broken, like the first morning.
Blackbird has spoken, like the first day."

He had always liked the third verse:

"Mine is the sunlight, mine is the morning
Born of the one light, Eden saw play.

*Praise with elation, praise every morning
God's recreation of the new day."*

Cat Stevens made the song famous. It was a story of Biblical creation, with echoes of Adam and Eve in the Garden of Eden, watching their first day.

Matthew knew peace and joy when his band played "Morning Has Broken." But he played with drugs, thinking it was innocent fun. Demons ensnared him, and joy was no more.

Now joy was back. He couldn't believe God had given him a second chance. Every day was a gift from God, a gift he didn't deserve. He didn't know how God had created the universe or stuff like that, who cared? He didn't need math or physics. God's creation was before him, like the first day.

Matthew placed the coffee mug down and bowed his head to pray. He prayed for his father. He prayed for victory against the demons of drugs. He prayed he would spend the rest of his life with Imani.

He deposited his empty mug on the counter and went to make coffee for Imani. At the machine's final beep, he walked into the bedroom and placed the fresh mug quietly on the nightstand beside her. Slowly, slowly, Imani began to wake, the scent of the coffee rousing her from deep sleep. She opened her eyes, saw the mug on the stand, and smiled at Matthew in thanks. Morning had broken, like the first day.

Matthew smiled back at her. "Get dressed, my love. There's something I want to show you."

Imani put on the thrift store dress, and Matthew led her down the back stairs to the kitchen and out the side door. They walked into an enormous glass conservatory, complete with a glass ceiling, in high English style. It was the largest conservatory the contractor had ever built. To Imani, it resembled a church made of glass but without a steeple, although the center part of the ceiling was raised. The symmetries of the glass, and the curved gothic arches between glass plates, were wonders she had never seen, never imagined. It was warm, mostly from the heated floor and vents but also from the winter sun now reflecting on the glass. There were flowers and shrubs and small trees in every direction. It was a veritable Garden of Eden, or so it seemed, and Imani was overwhelmed by the colors, the scents, and the warmth as she listened to the sound of babbling water.

Mary had been inconsolable after Matthew went to jail. She broke, and John had made the heartbreaking decision to admit her to a psychiatric hospital. He then went into action to "fix" the problem. The kids' bedrooms were replaced with designer showrooms. The English conservatory was added as a sanctuary for Mary. John told the architect to spare no expense and build a sanctuary of indescribable beauty, a sanctuary of light, warmth, and water. An army of botanists was tasked with recreating Eden. Italian sculptors were commissioned to cut the fountain and reproduce, using marble from the original quarries, two of the greatest masterpieces of the Renaissance, cut even larger than the originals. The sculptors used computers and lasers to make near-perfect copies. John paid top dollar and more. Four months later, just before Mary came home, it was finished.

Imani followed Matthew down the curved path, and the white marble fountain came into view. It was twelve feet high, cut in Palazzo style, with water cascading down four intricately carved tiers of bowls. Matthew took Imani gently by the arm and put his finger to his lips. He pointed to the back of the conservatory, to the East side, closest to the pond. There was a white marble bench, and Mary was kneeling before it, facing the rising sun. She was praying and crying silent tears.

Imani knew what to do. She took Matthew's hand and led him to the bench, where they joined Mary in prayer.

***

Matthew had never been in the conservatory. His mother had described it on their first phone call after his overdose. She had texted him pictures. She told him it helped her find her way back from darkness.

Matthew hadn't expected his mother to be there on Christmas morning. When they ended their prayers and all rose to their feet, Mary told him his father wasn't able to get out of bed. He was in great pain and breathing heavily, yet refused to let Mary call a doctor. If he was going to die, his father had said, he was going to die then and there in his own bed. He wasn't going to be hooked up to tubes and monitors like a lab rat. Matthew hugged his mother. He knew how much she loved his father and that she would do as he asked. No doctors would be called. He squeezed his mother tight and said, "I'm sorry, Mom." There was nothing either of them could do.

Matthew took Imani's hand. "Come," he said. "There's more." While his mother headed to the kitchen, he led Imani down another curved path, to the north side of the conservatory. There it was, in marvelous Carrera marble. A near-perfect copy of one of the most famous sculptures the world has ever known, cut larger than the original. It was Michelangelo's Pieta, the "Pity," except it was ten feet high. Mother Mary was holding her crucified son in her lap.

"Do you know what this is?"

"Yes," answered Imani, her voice soft with reverence, "but I didn't know it was so big."

"That's because it's not—the original, I mean. That's my father for you. Everything's larger than life. The original in the Vatican is 5 1/2 feet high. This looks about ten feet."

They stood in silence. Michelangelo's genius had turned stone into human flesh and folds of cloth. Now, it was Matthew's turn to be the teacher. He told Imani the original was over 500 years old. He pointed out that Mary was depicted as a young woman, perhaps eighteen or so, with a calm, serene face. He explained that Michelangelo carved her like that to portray her purity and the wonder of her virgin birth. Jesus was depicted as he would have looked at thirty-three years old after he was crucified. [26] Matthew pointed out the marks on Jesus's hands and side from the crucifixion.

Matthew took Imani's hand again. "Follow me," he said, excited to show her more. "The best is yet to come." He led her back past the fountain, this time to the south side of the conservatory. There was another statute, larger than life, and perhaps, just perhaps, even more astonishing.

Imani gasped. "That's David!" "Yes."

"He's pulling back his slingshot! Oh my God! Look at the anger! The strength! The grit! He's about to kill Goliath!"

Matthew told her Michelangelo also carved a famous statue of David, but this one by Bernini was the best. Michelangelo's David was a calm, motionless young man without obvious Jewish features, but Bernini carved him with a classic Jewish nose. Looking at the three- dimensional glory of the stone before him, instead of an image on a computer screen, Matthew saw skill beyond words. Bernini breathed life into the stone. David was reaching back, twisting violently, muscles flexing, anger exploding, ready to kill the man who had defied the living God.

Imani's eyes filled with tears. She took Matthew's arm and leaned into him. She told Matthew he should be proud of his father. His father had done all of this, built the conservatory with gardens and sculptures, because of love—to save the woman he loved.

They held hands and walked back to the kitchen. The sun was up. They didn't need to say it. They were wondering if John would be alive when the sun went down.

CHAPTER TWENTY-SEVEN

The bottom of the Christmas tree was overflowing with presents, and all but two were for Rebecca. Mary detested Madison Avenue for turning a sacred holiday into a feeding frenzy. Nobody argued with Iron Mary on something like that. Ashley had asked if they could bring Rebecca's presents, and Mary granted an exception. No need to ruin her granddaughter's Christmas. But Santa Claus was not allowed in the building.

They sat in the family room with coffees and pastries as Rebecca dug in. There were periodic exclamations of "that's so cute," "I love that," "where did you get it," "what a great gift," "you're going to like that," "how fun," and other similar inanities. When it was over, Rebecca tallied the take. She got 12 colored pencils, 11 books on science, 10 hair bows, nine books on history, eight graphic novels, seven designer outfits, six puzzles puzzling. FIVE MATH BOOKS. Four cutesy coats, three pairs of shoes, two cashmere sweaters, and a primer on introductory Mandarin.

After the presents, Mary got up and headed to the kitchen. Rebecca joined her a few minutes later..

"Did you like your presents?" asked Mary as her granddaughter hugged her.

"Another year without a baby brother." Mary hugged her back, laughing at Rebecca's cheek.

Mary began preparing her traditional Christmas feast. John loved Thanksgiving dinner, and Mary always made it again at Christmas. A large turkey with sausage stuffing was on the first island, waiting for the oven to warm up. There would be green beans with almonds, mashed potatoes, gravy, lima beans, warm bread, and warm rolls. Ashley was her sous chef. Matthew started on his contribution, cranberries in red wine—a dish with powerful flavors. The simplicity of the recipe matched his culinary skills.

Mark announced his contribution would be choosing the wine. He spent half an hour in the wine cellar and emerged with two bottles of wine, a Grand Cru French Bordeaux and a rare Napa Cabernet. Mark and his father had a tradition of picking wines and blind tasting with guests to vote on which was the best. Mark was not about to let a detail like his father's impending death interfere with tradition. He went to the cabinet in the dining room that showcased his father's collection of crystal decanters and brought back two. He opened the bottles and poured them into the decanters to let the wines breathe.

Tiffany and Imani set the table, under Mary's strict supervision, of course, with Royal Doulton china and Waterford crystal. The dining room was huge, but they didn't add leaves to expand the table. With the snow, Mary doubted any of the invited guests would show up.

John did not come out of the bedroom. Mary went back periodically to check on him; he refused to let anyone else see him.

The house was playing carols by the Tabernacle Choir. But the joy of Christmas was under the shadow of death.

Suddenly, the music stopped, and the house announced. "A large snowplow is at the front gate."

***

It was the Very High and Very Right Reverend John Jacob Pierpont Dingledorfer the Fourth. The local priest couldn't get out his front door, much less to John and Mary's estate. But he heard Mary's plea. Urgent calls were made, and one thing leads to another. The Very High and Very Right Reverend Dingledorfer was in charge of fundraising for the Holier Than Thou branch of the One Size Fits All Church. Not exactly connected to the local church, but, again, one thing leads to another, and the best they could get under the circumstances. When "Fourth"—the Right Reverend found 'JJ' inappropriately familiar but did allow close friends and prospective donors to call him Fourth—heard that billionaire John was dying, he felt an intense spiritual call to share the faith. He, too, made calls, and again one thing leads to another, and he eventually found himself, with the help of a generous gift of a case of whiskey to a friend of a friend of a friend, sitting in the passenger seat of a large snowplow. The driver, a fairly untrusting chap, insisted on quality tasting before and during the drive, but since they were in a large snowplow the occasional swerving didn't cause too much trouble, other than a few mailboxes that should never have been that close to the road.

Fourth's father, who everyone called Third, was a natural in ladies' underwear, in more ways than one. The success of his lingerie company allowed young Fourth to pursue his academic and theological dreams, which included a seven-year odyssey starting at Princeton and culminating in a mail-order degree in nonjudgmental psychology. Fourth then did a two-year internship at his father's golf club but sorely missed the academic life, and somehow—no one was quite sure how, but there were rumors of a connection between Third and the Dean—was admitted into Yale Divinity School. This time, after applying himself a trifle more, Fourth emerged with a degree in relativistic theology after only five years.

Mary went to the front door as Matthew ran to grab a snow shovel. Looking through a window, Mary saw a stout man step out of the passenger side of the snowplow. He was dressed in bright yellow foul weather gear, pants, and a matching jacket of the type worn by sailors in rough weather. The man lumbered painfully through snow above his waist. Matthew was shoveling snow off the front steps as the stranger finally made it.

"Hello," the man huffed. He took a moment to catch his breath. "Is this the Arnold residence?"

"Yes," said Mary. "And who might you be?"

"The Right Reverend Dingledorfer from the One Size Fits All Church at your service. I heard your husband was sick. May I come in?"

How strange, thought Mary. You never know how prayers are going to be answered.

The Very High and Very Right Reverend entered the foyer. He shed his boots, gloves, scarf, hat, jacket, and protective pants. He took a pair of leather loafers out of a pocket and slipped them on. He turned towards Mary, still breathing heavily. He was a little over five feet tall, about half of that wide, and clearly not a man who believed in wasting time at the gym. He now wore a white shirt, dark brown pants held up by yellow suspenders, and a bright pink bow tie.

The Very High and Very Right Reverend was a jolly fellow. He gave them his titles and credentials, making sure to mention Princeton and Yale Divinity School. He granted them permission to call him "Fourth." He oozed charm distilled from a lifetime of cocktail parties. He told Mary calls had gone out, and once he heard about John's condition, he had to come. And how miraculous he found a snowplow to deliver him. He explained to Mary it didn't matter what manner of religion she practiced, because the One Size Fits All Church was open to all, as inclusive as inclusive could be, of all faiths, all beliefs, and even to people who really didn't believe in anything, as long as they were willing to join together to spread happiness and joy. He began to explain the programs and benefits of the One Size Fits All Church, including the new combination brewpub and bowling alley.

Mary cut him off before he could continue his rant. "Would you like to see John?"

"Oh yes, yes, of course," said Fourth. "If you want me to and if I'm not going to bother him. I certainly wouldn't want to bother him, not in the slightest."

Mary led him back. Fourth's head twisted from side to side as he took in the mansion's size and opulence. "Nice house you've got here."

John's eyes were closed when they entered. Mary walked over and took his hand.

"John," she whispered. "Are you awake? You have a visitor."

John stirred, opened his eyes a little, and looked at her. Mary introduced Fourth, who began to tell John who he was and all about the One Size Fits All Church.

John's voice was weak but clear. "Get him out of here."

CHAPTER TWENTY-EIGHT

Mary didn't know what to think. She felt bad she had woken John when he had asked to see no one but her. But she also felt sorry for Fourth. The guy seemed pleasant enough, not exactly a Billy Graham or Mother Teresa, but he had taken the trouble to be there. Out of guilt, and sheer politeness, she invited him to stay for dinner. There was plenty of food to go around.

Fourth was delighted to accept. After being introduced to the rest of the family, he was also delighted to be shown where the liquor was kept. As Mary headed off to check on the turkey and Fourth began to make himself a vodka martini, a low rumbling noise came from the back yard.

When the noise only grew louder, Mary went to peer out the window and saw a snowmobile emerging from the woods. Its driver wore a full-length coat of thick Russian sable, a matching Russian-style fur hat, and ski goggles. *Well, whaddaya know*, thought Mary, smiling to herself. It was Svetlana Whitehead from three estates over. Mary hadn't seen her for quite some time. *What a pleasant surprise*, she thought as she headed to the back door on the side of the family room to let her friend in.

But the Svetlana standing at the back door was not the Svetlana Mary remembered. The old Svetlana had been a professor of biology at Columbia University but left that position three years ago to head up the National Academy for the Glory of Science. She traveled constantly to alert everyone to the dangers of believing the Bible was true. She thought of herself as Darwin's defender; others called her the terminator. If a teacher dared suggest science supported the Bible, Svetlana was there, and the poor teacher almost always found herself or himself quickly unemployed. She was a crusader. Her religion was Darwin, and she was a fanatic.

Her crusade did not help her marriage. Her lonely husband developed other hobbies, including the study of gymnastic positions with his twenty-seven-year-old physical trainer. He and Buffy were now sunbathing on his 200 foot yacht somewhere in the area of St. Barts. Svetlana fervently hoped the two of them would be swept overboard by a rogue wave and drowned, but she was smart enough to know that was not likely to happen. From the telephoto pictures her private detective had sent back, it was apparent the clever young woman had two roundish floatation devices implanted in her chest, doubtless to forestall that very danger.

Mary was shocked by the new Svetlana. The designer clothes and dripping diamonds were not new (Plain Mary was one of the few women in that part of town who did not favor the "I'm-richer-than-you" style of dress). What was new was a facelift pulling back Svetlana's skin and turning her into something reminiscent of a horror movie. Also new was her hairstyle. The once cute short blonde hair was now, through the miracle of modern hair spray, something that resembled nothing so much as a football with a space cut out of the bottom so it could be squashed on top of her head.

Svetlana told them the storm had prevented her twenty-years-younger Italian boyfriend, whom she met skiing in Switzerland, from flying in for the holidays, and so she was delighted to accept Mary's invitation. Mary introduced Svetlana to everyone before returning to the kitchen to work on dinner. In the background, the house was playing soft music.

Suddenly, the music got louder:

"I want a baby brother for Christmas
Dear Santa, won't you bring one to me
How I'd like to find a baby brother
In a bundle underneath the Christmas tree
I want a baby brother for Christmas
Dear Santa Claus, that's all I ask of you
If you can't take my order for a baby boy
Then I guess a baby sister will do!"

Everyone looked at Mark, who shrugged, said he had no idea why the algorithm had chosen that song. He had never heard it and didn't know why the music had gotten louder. It was probably one of the glitches his engineers were working on. Mary glanced at her angelic granddaughter lying on the floor and playing with one of her new puzzles. Was it possible?

Mary went again to check on John. His eyes were closed and he was barely breathing. She knelt and prayed beside him. Tears ran down her face. After a while, she stood up, took a deep breath, and dried her tears. She put on her New England Yankee face, the stoicism of her forebears, and headed back to the kitchen.

Under Mary's instructions, Imani and Tiffany reset the table. Wonderful smells filled the house. But joy was darkened by the shadow of death.

As Mary worked on dinner, a loud knock on the front door echoed through to the kitchen. The final guest had arrived.

CHAPTER TWENTY-NINE

Mary had company this time as she walked to the front door. Everyone was curious. The snow was three feet and then some. Who else could have made it?

Mary looked out the window but saw nothing. There was no vehicle. Thick shrubs blocked the front steps.

She opened the door to find a tall, thin young man. He wore a rough coat made from camel hair. He had neither hat nor gloves. His long hair was pulled back in a bun and his chin covered in uneven stubble. He held a Bible in one hand and a plain brown paper bag in the other. "Hello," he said, "I've come to pray for John."

His voice was kind. Mary asked if her local priest had sent him.

"I heard the call," was all the stranger said. "I've come to pray for John."

It sounded mysterious, but John needed all the prayers he could get, and Mary wasn't about to leave a polite young man standing in the cold. She invited him in. He said his name was Eli Wu. He took off his boots and handed Mary the paper bag. Mary peeked into it and found a jug of red wine. She looked quizzically at the young man.

"This wine is for you and your household," said Eli. "Drink it in remembrance of Jesus." Then he smiled and asked if he could see John and pray for him.

Mary wasn't sure if taking the man to her husband was a good idea. It would be against John's last wishes, and John hadn't wasted any time kicking the Right Reverend out. But there was something about Eli, a gentleness and a sense of peace, something that made her say yes. Eli followed Mary to John's deathbed with his head bent low. Mary stood at the foot of the bed as Eli walked in. She watched as he stopped, smiling kindly at John. He took a deep breath and closed his eyes in silent prayer. Then Eli walked around the bed and knelt next to the dying man.

John was barely breathing; he did not move or open his eyes.

Mary sensed the presence of God. Eli put his Bible on the bed, clasped his hands, bent his head, and recited from memory the prayer of David, the 23rd Psalm:

"The LORD is my shepherd; I shall not want.
He makes me lie down in green pastures.
He leads me beside still waters.
He restores my soul.
He leads me in paths of righteousness
for his name's sake.
Even though I walk through the
valley of the shadow of death,
I will fear no evil, for you are with me;

your rod and your staff, they comfort me.
You prepare a table before me
in the presence of my enemies;
you anoint my head with oil;
my cup overflows.
Surely goodness and mercy shall follow me
all the days of my life,
and I shall dwell in the house of the LORD forever."

Eli lifted his head and gazed at John. He took John's hand. Then Mary saw something. Something so fantastic her mind couldn't comprehend it was happening, something so real her eyes could not deny it. A ball of light surrounded John and Eli, and Eli's face glowed with compassion. Mary was both awed and scared.

Eli spoke. "I thank you, Father, for you have heard my prayer. In the name of Jesus, rise."

Immediately, John sat up in his bed. Mary truly, absolutely, could not believe her eyes. She watched in disbelief as John turned and put his feet on the floor. He shook his shoulders and moved his head slowly from side to side as he regained his senses.

And then, John, his voice clear and strong, asked, "Do I smell turkey?"

CHAPTER THIRTY

ary didn't know what to think. There were no words. She and Eli tried to help John walk out of the room, but John pushed them away, insisting he was okay on his own. He went off to clean up for dinner while Mary walked to the kitchen in shock.

Eli followed and thanked her for letting him pray for John. He began to pull on his boots to leave.

Mary jumped between Eli and the door. "Please stay!" she said. "You must stay." From somewhere, she found words. "In the name of Jesus, please stay for dinner and pray for all of us."

Eli looked at her kindly. "At the name of Jesus, every knee should bow, of those in heaven, and those on earth, and of those under the earth. [27] Because you ask in the name of Jesus, I will stay."

A bewildered but thankful Mary announced to an astonished household that John was feeling better and was getting ready to join them. She introduced Eli awkwardly because she didn't know anything about him. Eli shook hands and said hello. Mary explained that when Eli told John to get out of bed, he did exactly that. She said it was a miracle. She didn't mention the ball of light. It was too fantastic, and she didn't want to sound crazy.

Mary took note of Eli's simple clothes. She noticed a dark, uneven spot on the back of his hand and his palm, as if he had been badly wounded.

"What happened?" asked Mary, her gaze on his injuries. "Were you injured?"

Eli shook his head. "An occupational hazard." Imani was curious. "Are you a stigmata, like St.

Francis of Assisi and others over the centuries? Do you carry the mark of Christ?"

Eli smiled. "We all carry in our body the death of Jesus, so that the life of Jesus may also be revealed in our body." [28]

Silence followed these mysterious words. Mary didn't know what to say. She asked Mark to get a third decanter for the wine Eli brought. She checked on the turkey and busied herself with side dishes. Mary was in shock. Something was going on, but she didn't know what it was.

CHAPTER THIRTY-ONE

Mark was surprised to see the three new visitors and hear about the reported improvement in his father. He did as his mother asked as he tried to take it all in. He went to the dining room and brought back a third decanter, sneering as he took Eli's wine out of the paper bag. The wine was in a cheap jug with a handle, the type of bottom- shelf wine no self-respecting person would buy. It was an insult to show up at a house like this with wine like that. He was worried someone might think it was a wine he picked out, so he wrote "Eli's wine" on a small scrap of paper and taped it to the decanter. *That should prevent any confusion*, he thought.

Across the room, Fourth was giving Svetlana his full attention, talking loudly about the One Size Fits All Church. He didn't seem to mind if anyone overheard, exclaiming that, because you didn't have to believe in anything or change anything about yourself, the One Size Fits All Church was growing rapidly. It was about feeling good and having a good time. In response to a question Mark couldn't hear, the reverend stated that yes, of course, the One Size Fits All Church embraced all manner and practices of the joys of human sexuality and was a leader in the beyond binary movement. Svetlana seemed to like that answer.

Mark needed to talk to Ashley and Hajid. What did they think of these twists? Did the new guests have something to do with his father's recovery, if what his mother said was true? Did the new events help or hurt the strategy they had "agreed" upon? And what about Plan C-cret, his absolutely brilliant idea of convincing his father that Mark was the only person capable of carrying out his wishes? Any changes to be made there? Maybe not. His father's recovery, if it was true, was great news. The more time he had to talk to his father—privately— the better. Bottom line, his secret plan was looking good.

They couldn't disappear again to the wine cellar, so they met in a corner of the family room.

"This is fantastic!" whispered Ashley. "Do you know who Svetlana is? She heads up the National Academy for the Glory of Science, and she travels around the country to promote science. Hajid and I met her at a dinner once. She praised MIT for having 'Darwin' carved in large letters above the columns on our main building. [29] This is fantastic. Svetlana can explain why science, the power of human intellect, is the hope of the world, the *only* hope of the world. She can explain natural selection and common descent, Darwin's theory that all creatures are descended from a single ancestor that, through the power of natural selection acting over billions of years, became all of the forms of life, plants, and animals we see today. Frankly, I'm not sure I believe it, I've seen studies showing the theory can't be true, but Svetlana sure does, and she will argue it forcefully. She will argue science has replaced the Bible and that human intellect will save us. She's perfect! If there is any suggestion the Bible is true, such as some fable of Adam and Eve in some garden, Svetlana will crush it. She will be our champion."

"And this 'Fourth,' or whatever he calls himself, is a clown," said Hajid softly. "He even looks like a clown. He's a caricature of religion. Your dad is going to be repulsed. He's not going to want to give money to any God that has Fourth as a foot soldier."

"What about Eli?" asked Mark.

"Not sure," admitted Ashley. "He seems to be simple, probably without education. Mom thinks he could be a handyman at her church, because it looks like he shot himself with a nail gun. I think he's stupid." Ashley smiled. "Don't worry about Eli."

Mark needed to learn more about Eli. "Let's talk to the guy. That can't hurt."

With that, Mark strolled over to Eli and offered him a drink.

"A glass of water would be great."

When Mark brought the water, he invited Eli to join them in the corner of the family room, where he began the interrogation.

"Can it be true that our father is doing better?"

Eli smiled a little and responded slowly and softly. "I hope so. Your father was doing better when we left."

Ashley was quick to jump in. "How did that happen? What did you do? What did you say?"

Eli responded just as slowly as the first time. *Maybe Ashley was right*, Mark thought. *Maybe this guy wasn't very smart. Perhaps he was, as they say, mentally challenged?*

"I just prayed. I prayed for your father. I asked Jesus to save him."

Ashley went for the kill. "What's your background? Did you go to college?"

Again, Eli responded, very slowly and softly. "I went to technical school for a year and a half. But I dropped out."

"Why was that?" pressed Ashley. "If you don't mind my asking?"

Eli paused and looked innocently at Ashley. "I just wanted to pray. Now I pray, and I wander. I get work where I can."

Mark thought that was enough. They had what they needed—they were dealing with a simpleton. No need to embarrass the poor kid more.

"Thanks," said Mark. "Do you mind if the three of us continue our conversation? Would you like more water?"

Eli got up and backed away. "There's no need. I can get it myself. God bless all of you."

As Eli turned slowly and walked away, Mark heard Ashley chuckle. "Nothing to worry about there. Not the sharpest blade in the drawer. He's not going to be a problem. He probably never heard of Darwin's theory, much less the new evidence against it."

"And don't forget your father is suffering, he's in pain," said Hajid. "Only science can end suffering. Bottom line, Plan A is looking good."

But Mark wasn't so sure. For one thing, he didn't trust Ashley and Hajid, although he couldn't think of a reason why they would lie to him, they seemed sincere. But also, more importantly, there was something about Eli. Something mysterious he couldn't put his finger on.

"I like that," said Mark. But inwardly, he didn't like it at all. Now he didn't want his father to sour on religion.

Now he wanted his father to realize Mark was the one to carry out his wishes. *And what the heck*, thought Mark. *I may actually do that, carry some of them out after he dies, as long as it doesn't cost too much money.*

CHAPTER THIRTY-TWO

John was feeling good. Was it a delayed reaction from the pills Mary had forced down his throat? He didn't know. He took a shower, shaved, and dressed in clean pants, a dress shirt, dress socks, shoes, and a navy-blue blazer. The shadow of death had passed, at least for a while. It might come back, it might come back soon, but nothing was to be gained by worrying. It was Christmas. He was going to spend it with family. Joy was in the world.

He walked in as Mary was basting the turkey. There were gasps and cheers at his recovery, and he was greeted warmly by his family. John sat at the first island to be close to Mary as she and Ashley prepared the side dishes. Rebecca came around and hugged him. John said hello to Svetlana and dutifully acknowledged the presence of the chubby man with the yellow suspenders and pink bowtie. After a while, Mark, Ashley, and Hajid left to continue their conversation. Matthew asked Eli to help him add a leaf to the table and Mary asked Tiffany and Imani to add a setting for Eli and extra glasses around the table for Eli's wine.

John was feeling good. His head was clear, his senses were sharp, and his pain was gone. He sat dazzled by the wonder of family joy at Christmas, the sounds, the smells, the colors. He was astonished, shocked, by the wonder of existence and the gift of life.

There was one thing he had to do. He had to give Mary her Christmas present. He waited until the side dishes were prepared. Then he turned to Mary. "Come, sweetheart. I want to give you your Christmas present."

They walked into the family room, and everyone followed. John and Mary sat together on one of the two couches that flanked the fireplace. Beneath the tree were two lonely presents. John stood up, walked to the tree, picked up a small package, and handed it to Mary.

Mary was suspicious. She looked at her gift with amusement. To start with, the wrapping was too neat for John to have wrapped it. The small size, a cube about three inches on a side, was also suspicious. It looked like jewelry and she had lectured John over the years about the importance of not making Christmas all about gifts. She had laid down the law, they could each give each other one gift only, and it had to be inexpensive.

She slowly unwrapped her present and, sure enough, inside was a jewelry box. She looked sternly at John. "I hope you remembered our agreement." She slowly opened the box to reveal a heart-shaped diamond. Not a super big one but not a super small one either, maybe one carat, maybe a little more. It sparkled in the light of the fire and was set in a simple silver chain. It was beautiful and Mary was touched. She turned to John for an explanation.

"Forgive me. I wanted to give you something to remember me by."

Mary hugged him tightly. Her eyes glistened as she kissed him on the cheek and whispered thank you into his ear. She fetched the last present and handed it to John.

When the final layer of wrapping paper was removed, John's face lit up. It was a book Mary had spent months putting together, a picture album of their life. There were pictures of them when they first met—when their friends had fixed them up, pictures of when they got engaged, of their wedding and the simple reception in the parish hall, pictures of their first tiny apartment in the city, of the children when they were small, of family holidays and family vacations, more pictures of the kids growing up, more of holidays and vacations, pictures of the kids when they were older, and pictures of Rebecca growing up. It was bound in leather and gave off that singular scent that only comes from good leather. They turned the pages and shared memories.

There was still time before the turkey was done. Mary remembered Matthew and Imani had rehearsed a song for Imani's church. Imani had a powerful voice. She asked them to sing it.

CHAPTER THIRTY-THREE

Imani was uncomfortable. It was bad enough being jerked out of her world at Christmas, but to be sitting in this house, with these people? All her life she had resented, often hated, sometimes feared, people with money. And these people had more money than God. Not really, but seriously, how could anybody have this much money? The woman with the weird hairdo was scary and vulgar; nobody should have or wear that many diamonds. The fat guy with the bowtie was a joke. She had on her thrift store dress, which she was wearing for the second day in a row. And to state the obvious, she was black and they were white. Not that white people were bad, but these people sure were different.

She was surprised she liked Tiffany, though. Despite appearances, Tiffany was at least genuine. And John and especially Mary were warm and welcoming, and obviously wanted her to be there, and she knew they knew she was the rock for their troubled son. And she loved that son. She really did; she had fallen hard for Matthew. But it was strange, and it was uncomfortable. And now they wanted her to sing? Really? At least Matthew would sing with her, wouldn't he?

But Matthew had other ideas. "Please sing 'Amazing Grace,'" he asked. "You are better singing solo. Please do this as a gift to all of us."

Imani had no choice. She nodded, and Matthew stood up. "Let me tell you a little about Imani. Her name means 'faith' in Swahili, in the part of Africa where her ancestors came from. She is the star of our church choir. This Christmas, we were going to sing 'Amazing Grace' together, as a duet. But I want you to hear her voice alone."

Matthew continued. "'Amazing Grace' is perfect for Imani. It is the most popular Christian hymn, and perhaps the most popular song, if you look at how many times it has been recorded and sung, in all of history. It's the unofficial anthem of the civil rights movement. It was written by an English sailor, a man named John Newton, a man profoundly troubled—so profane other sailors were put off by his language. The twists of his life are incredible. He was pressed into service—literally grabbed off the streets and forced to be a sailor in the English Navy. They traded him to a slave ship, and he was almost starved to death, shackled next to the slaves, and later was himself enslaved. He became the captain of a slave ship, a ship that transported shackled African prisoners to the Americas to be sold as slaves if they survived the hell of the voyage. Then there was a storm, a storm where he thought he would die, a storm where a colleague next to him was swept away, and he called out, 'Lord have mercy upon me.' Jesus then began to use John Newton. He had had only two years of poor schooling, but he taught himself Greek to read the New Testament and Hebrew to read the Old Testament. He read the Bible, and it changed him. He became a priest and, 250 years ago, penned the words that have resonated for millions as a message that, through the pure and undeserved grace of God, even 'a wretch like me' can be saved.

"And it's true," said Matthew, "God's grace *is* amazing. You see, that's the key. It's all about God's grace, the undeserved mercy of God. At the end of his life, as he was blind and fading, John Newton was asked if he remembered something. He responded, 'Sir, I know only two things, that I am a great sinner and that Jesus Christ is a great savior.'

"And here's the kicker. God used John Newton to end the slave trade. His powerful preaching, his straight- from-the-heart story of salvation, and his first-hand knowledge of the horrors of slavery, changed the world. The story of John Newton is the story that God can redeem and use even the worst of us."

With help from Rebecca and some grudging help from Mark, the music system was set to play background music for Imani.

Imani stood up. She started tentatively:

"Amazing grace, how sweet the sound
That saved a wretch like me.
I once was lost, but now I am found
Was blind, but now I see."

Astonishment lit on every face. If one thing was true, it was that Imani knew how to sing. Now she was not uncomfortable. Now she was radiating, at her best, and her voice was powerful, rich, intense, and pure:

"T'was grace that taught my heart to fear
And grace my fears relieved.
How precious did that grace appear
The hour I first believed."

Perhaps there never was, before or since, a more wondrous song, sung by so powerful a voice. Imani put her soul into it. Her voice resonated off the family room walls, with rich harmonics, flowing like a river, with a deepness and beauty that was out of this world. Imani came to the last verse:

"Amazing grace, how sweet the sound
That saved a wretch like me.
I once was lost, but now I am found
Was blind, but now I see."

John was in tears. He couldn't control himself. Mary squeezed his hand. It was magic. Never was there such a moment, as when Imani sang Amazing Grace.

CHAPTER THIRTY-FOUR

The music shook Hajid. Hard enough to crack, for a few moments, his shell of intellectual snobbery. Hard enough so that for a few moments, he was no longer a tenured professor at Harvard, no longer a person of international renown. He was a person who had seen Heaven.

They were sitting in the family room, some on the two couches and others in chairs around the fireplace. The house was awaiting instructions as to the music, so it was quiet, the room filled with gentle conversations. Imani was sitting near him, so Hajid had to ask.

"Your voice shook me. More than I expected, more than I can explain. It was beyond beautiful. You took me to another world. I had a glimpse of Heaven. It was amazing! I can't say it enough. You were amazing."

Hajid was the last person Imani had expected to address her like this. "Thank you" was all she managed to say.

Hajid took a breath. Imani's voice had banished his demon of intellectual superiority. Free of the demon, Hajid sat in grace and awe. An inexplicable joy came over him. It was the joy of being able to see humanity in a different person. He was not in competition with Imani, far from it. He needed to learn from her. For now, he wasn't a professor. He was a person seeking truth.

"No, thank you. I need to apologize for last night. You're right. We have no clue where the laws of physics come from, or why they are fine-tuned, set just right for life. No clue. And we still haven't processed the discovery, close to a century ago, that the universe had a beginning. We're in denial of a fact that humbles all mankind. The multiverse is nonsense that doesn't work."

Imani was surprised to hear the words. *Hajid was human after all*, she thought. He continued.

"I'll be honest. I can see the universe was created and designed. I see there has to be a God, a mind, who created it all. But that doesn't help me. It doesn't change my life. It's an intellectual assent to a truth claim, nothing more." Imani interrupted. "Even demons know God is real.

They recognized Jesus immediately. Just saying God is real gets you nowhere. If it doesn't change your life, you are lost."

"I wish I had your faith," lamented Hajid. "Your faith, your serenity, your peace. It's beyond price. But here's my problem. Why should God care about me? For that, doesn't it come down to Jesus? You believe Jesus rose from the dead. That violates the laws of physics. Science, and physics, are part of my being. How can a scientist believe in Jesus?"

Hajid could see Imani didn't know what to say. He repeated.

"I won't be a fool. How can I believe Jesus rose from the dead?"

Again silence. Imani did not respond. Then Eli piped-up.

"That's a great question. May I answer?"

Hajid didn't have a clue why Eli might have some-thing to contribute, but he was still under the spell of Imani's voice, free of his demon, and he wanted to learn. "Yes, please."

Eli asked John for permission to sit next to him on the other couch. John nodded yes, and Eli sat down facing Hajid. Eli was not the sort of person you would expect to be discussing physics across from a Harvard professor. Nor was he the sort of person you would expect to be sitting on a couch next to one of the richest persons in the world. But there was something about Eli, and, who knows why, at that moment it seemed natural.

"I see it the other way," said Eli. "To me, any scientist, and any rational person willing to look at the evidence, has to admit, whether they want to or not, that Jesus rose from the dead."

Hajid blinked. That was a surprise. He was sitting across from someone who saw things another way. Hajid let him continue.

"We have four eyewitness accounts. Four gospels. They mesh like eyewitness accounts do today. Minor differences, same powerful story: the story of a messiah who performed miracles, was crucified, and rose from the dead. You can tell the gospels are not made up by details, details you would not put in if you were making up a story. Details such as the first people to find the tomb empty were women. Back then, women were not allowed to testify in court, their testimony was considered unreliable. Nobody making up a story at that time would have their first witnesses be women.

"Because of Roman records, and other evidence, we almost certainly know the very house where Jesus grew up, almost all scholars today, even atheist scholars, agree Jesus was a real person. They also agree he was crucified, a horrible form of torture, and died nailed to a wooden cross. The Roman soldiers risked death themselves if they let any person being crucified survive that torture. Scholars agree the tomb was empty that Sunday morning.

"Could someone have stolen the body?"

"No. The Romans rolled a large stone over the entrance, sealed it, and set an armed guard. These were trained career soldiers. They were instructed by Pontius Pilate, the Roman governor, to make the tomb 'as secure as you can.' [30] Suggestions like the soldiers walked away from their post, or slept through the stone being rolled away, defy common sense."

Eli continued. "So, Jesus was a real person, was crucified—nailed to a wooden cross and killed by torture—and his tomb was empty. Now, did he rise from the dead?

All of Christianity hangs on that question. Two thousand years ago, Paul wrote that if Jesus didn't rise from the dead, the Christian faith is empty and Christians are 'most to be pitied.'" [31]

"I agree," said Hajid. "What's your evidence?" "Overwhelming. Spectacular and overwhelming.

Again, four eyewitness accounts. Plus, other writings. Paul tells us that Jesus once appeared to over 500 people, and most of those people were still alive when he wrote that. [32] He couldn't have written that if it wasn't true. All of Jesus's disciples, except one who was exiled to an island off Turkey, were brutally murdered. They refused, even when facing horrible deaths, to deny what their eyes had seen, that Jesus rose from the dead. No one dies for a useless lie."

John interrupted. "I never thought of that. Perhaps for a good person or a good cause, a person might lay down their life, but no one would die to defend a made-up story, a useless lie that did not help them or their family. These guys had to have been there and seen the resurrection. They would not let themselves be tortured if it wasn't true."

Eli nodded and kept going. "All historians agree that within a few decades of Jesus's death, there were enormous numbers of Christians, including tens of thousands in Rome itself. They didn't become Christians for money; many of the first Christians gave away everything they had. They didn't become Christians to gain status, the

Romans hated and killed Christians; there were at least twelve waves of horrific persecution. Yet, Christianity spread like wildfire. Something was going on."

"I hear you," Hajid replied. "Lots of history, lots of facts, no other explanation. But you haven't answered my question. How could the laws of physics be violated?"

Eli sat back, relaxed, and smiled. "And I hear you. The answer is simple but hard for you to see. God thought the world into existence. God *is* existence. In the beginning was the Word, [33] the divine thoughts of God. We live in the mind of God. In him, we live and move and have our being. [34] God has total control of everything, every grain of sand, every atom on every star. He knows the stars by name, he knows the hairs on your head. In your language, the language of physics, he has absolute control of every quantum phase state. When God wants to think different thoughts, he does. God can change the laws of physics. He wrote them into the universe, and he can change them. He's God. God can do things we can't understand."

Hajid was unsure, and his face showed it. "I don't know. I still have problems. Let me ask you another question, a question that has always bothered me. If God made the universe, who made God?"

Eli leaned back and smiled. "Great question." He paused. "When we say the universe had a beginning, and I think you agree with that, it's now commonly accepted, it's not just space that had a beginning. Time itself had a beginning. You can't go back in time before God created the universe. Time didn't exist."

Eli looked intently at Hajid. "You see, our concept of causation, our concept of one thing leads to another, has two events separated by time. But outside time, outside this fishbowl of a universe, who knows how things are connected? Who knows how things come to be? It is not just stranger than we imagine, it is stranger than we *can* imagine.

"When you ask who made God, you are trying to look back to an imaginary moment before the universe was created. But you can't do that. Time didn't exist."

Hajid sat, mulling over what he was being told, and Eli continued. "Two things may help. First, as Einstein taught us, time, space, matter, and even energy are connected. The equations of general relativity are a work of indescribable beauty. You can't separate time from space, matter, and energy. When space, matter, and energy did not exist, time did not exist.

"Second, all of this comes from the Bible. The Bible tells us God is existence. On the mountain, when Moses asks God for his name, God says his name is 'I am.' God says, 'Tell them "I am" has sent you.' You see, God *is* existence. Nothing exists without God."

Eli smiled mischievously. "But I get it, it's strange. We want an answer we can wrap our minds around. We want a picture in our brains of why things exist, why there is something rather than nothing. But our brains are in the fishbowl, made of matter and energy, dwelling in time and space. We can't see outside the fishbowl. God just is. As for what's outside the fishbowl, we only know what God tells us."

Eli smiled kindly. "You don't realize you're in a fishbowl. And you're not the only one. Unless you let God into your soul, you will never be able to see, or even get a glimpse of, what lies beyond the fishbowl."

CHAPTER THIRTY-FIVE

Fourth was making his fourth martini. There were reasons for that. For one, the One Size Fits All Church was all about conservation of the earth's resources. There was no point letting vodka and vermouth go to waste. Fourth practiced what he preached. Second was Fourth's impeccable manners. Mary had instructed him to "help himself," and like any well- mannered guest, Fourth was obliged to abide. But most important was his mission, to cover all humanity under the umbrella of the One Size Fits All Church. Experience had taught him perseverance was key, and nothing helped him persevere quite as well as multiple martinis. To Fourth, and most of the blessed faithful of his church, drinking was a religious experience.

It was clear to Fourth, and more so with every martini, that these people needed his spiritual guidance. If one thing was clear, it was that none of them had gone to Yale Divinity School, much less managed to somehow graduate. The joys of the new spirituality were there for the taking, or the drinking, depending on one's preferred form of worship. Fourth could lead them.

Drink in hand, the Very High Reverend walked from the liquor cabinet back to the fireplace. Years of training allowed him to maintain a steady and stable gait. His educated ear alerted him the conversation had something to do with Jesus. Time to shine.

"If you don't mind," interjected Fourth, "I do think you're making it too complicated. There are things we don't know. But at the One Size Fits All Church, we focus on what we do know, the things that make people happy."

John was looking at him. It was not the kindest of looks, and it did not contain the respect and admiration due to someone who had briefly attended Princeton and somehow graduated from Yale.

"Exactly what is that?" asked John, one eyebrow raised.

Fourth smiled. One had to start somewhere with new converts.

"The human race," he said, "is on the verge of achieving an ancient dream. We can create Heaven on Earth. No longer do we need to be divided by our beliefs. No longer do we need to be divided by our practices. Together, we can build paradise. We have the technology. We can eliminate hunger. We can defeat disease. We can ease suffering. We can do it together." Fourth looked proud as he stated, "That is the mission of the One Size Fits All Church."

It was a brilliant speech, even if he did think so himself. Just then, Mary walked back to the family room. *How wonderful,* thought Fourth. *Another soul to save.*

"But what about Jesus?" asked Mary. "How does Jesus fit in?"

"Great question!" said Fourth, perhaps a little loudly, but sharing the faith was his passion. "I'm so glad you asked. The One Size Fits All Church has the utmost respect for Jesus. We really do. If Jesus makes you happy, we welcome you."

"What about people who are not sure about Jesus," challenged Hajid, "who have doubts?"

Perfect, thought Fourth. *What a joy to be given the opportunity to help the spiritually challenged.* "Absolutely no problem!" said Fourth, again a little loudly, but this was working out quite splendidly. They were pitching him softballs, and he was going to hit them out of the park. "At the One Size Fits All Church, we recognize Jesus is not for everyone. Not that we're against Jesus, he was probably a nice guy, and we do feel sad about what happened to him, quite a pity if you ask me, but we stand for the truth, and the truth is Jesus can be divisive. People say this about him, people say that, and the next thing you know, you have an argument. Wars have been fought. The guy can be a problem. We don't talk about Jesus. Oh no, we don't talk about Jesus." "And what if," joined in Svetlana, "and I hope I'm not offending anyone here, but I did teach science at Columbia, what if one is dedicated to science, and the power of human intellect, and knows, like any person should, that the material world is all there is, and that we need to be rational about it. Don't you think we should talk about science?"

"Oh yes!" exclaimed Fourth. "The One Size Fits All Church glories in human achievement! Marvelous stuff, your science, simply marvelous. Well done!"

"Some say that science, and human progress, were birthed by the cross, that only in the secure knowledge of a rational universe did people work together to figure out how God did it. I'm not sure," said Hajid, "but it smells of truth. What do you think?"

"Perfect example!" said Fourth. "Don't you see? Anytime Jesus gets near, you've got problems. Some say this, some say that, people argue, nobody's happy. We don't talk about Jesus."

"But aren't we falling apart?" asked Mary. "Suicide rates are up; mental health is collapsing. The Chinese government is a nightmare of state control, the Russian government threatens nuclear war. Good people in those countries and around the world are being crushed. Don't we need Jesus?"

"If it makes you happy, that's fine," allowed Fourth. "But the solution is to expand the One Size Fits All Church. When everyone is a member, there will be no need to fight."

Eli was looking at him. There was something about that, something about the way Eli looked at him, that gave Fourth pause. Fourth looked back at him just as Eli spoke. "Then tell me," requested Eli. "If you don't worship

the Creator of the universe, who or what do you worship?" Fourth smiled again. Another softball. "My dear
chap, isn't it obvious? At the One Size Fits All Church, we
worship ourselves."

CHAPTER THIRTY-SIX

It was the house that decided to end the discussion. Or so it seemed. Music began blaring out:

"And the load
Doesn't weigh me down at all,
He ain't heavy, he's my brother."

Then the last line played over and over, like a broken record. "He ain't heavy; he's my brother."

The lights, even on the Christmas tree, began to blink. At first, it was synchronized, everything off, everything on, but then things went off kilter, with not just different cycles but random patterns. The blinds started opening and closing. The TV started showing pictures of baby boys.

Everyone looked at Mark. He apologized profusely and ran for his laptop.

But Mary knew better. She looked at Rebecca, who was grinning from ear to ear. Mary laughed. "That's enough, Rebecca. Cut it out!"

***

It took a while for things to get back to normal. That's assuming things were normal to begin with. Which they weren't.

The turkey was close to done, so Mary went to the kitchen to put the side dishes in. She still had to make the gravy. It was normally John who made the gravy. It was their little tradition, and he was good at it. But Mary wasn't sure John was up for standing by the stove.

Meanwhile, John was proud of his granddaughter. A nine-year-old girl had taken down Mark's system. That was amazing. Where does intelligence come from?

John wanted to hear more from Eli. He might be scruffy, but he had presence. So much was happening, and John was trying his best to make sense of it. Maybe hearing more from Eli would help.

"Would you mind continuing our discussion?" asked John. What's this about outside the 'fishbowl,' outside our universe? What did you mean by that?"

Christmas dinner was almost ready. But the conversation was about to get serious.

CHAPTER THIRTY-SEVEN

Eli sat up straight. He tried to hide the smile on his lips, but the sparkle in his eyes gave it away. Everyone except Mary was in the family room. Mark and Svetlana looked bored; Fourth was nursing his fourth martini. Everyone else was paying close attention, even Rebecca.

"Worldview. That's key. What's your bottom line? Is our universe all there is, or could there be something more? Are there things unseen, things beyond, things hidden from our eyes and senses? Could there be things unseen? Is it possible our senses don't tell us everything about reality?"

Eli looked at Svetlana. "You say 'no.' You say this universe, the material world, what we can detect with our senses, is all there is, all there can ever be. But why? You can't prove that. But it is your rock-bottom, fundamental, non-negotiable assumption about reality.

"You see, we each have a worldview, a bag of starting assumptions. It's a lens that brings everything into focus. It helps us make sense of things. How our lens works and how we perceive reality depends on our starting assumptions, more than most of us realize."

Eli looked kindly at Svetlana. "You want to call your lens reason, but that is not its real name. Some call your lens naturalism, some call it materialism. Your starting assumption is that the material world, what you call 'nature,' is all there is. You don't think of that as an assumption, you think of it as absolute fact, the hard truth of reality. But it is an assumption."

Eli looked patiently at Hajid. "And science proves it is false! It contradicts quantum physics. Quantum physics exposes the lie of materialism. Quantum physics is the mathematics at the bottom, the foundation of the universe. It's the thoughts of God. Any scientist who studies quantum physics must admit the material world is not all, there is something more."

Hajid sat stunned. Eli kept talking, and now swept his gaze across Svetlana, Ashley, and Hajid.

"Your lens, your worldview, your materialism, your lens that filters out God has become the dominant worldview of our global culture. You teach it at your universities. It is an unspoken requirement to get ahead, to be a member of the club."

Ashley interrupted. "That's not true!"

Eli's expression turned sad. "Please. Be honest. How many professors do you allow to teach, or even present, the truth of the Bible, that God created the universe in seven days, that God created human beings and all life?"

"Serious scientists don't believe that!" insisted Ashley.

"But they do," said Eli softly and sadly. "Thousands. Thousands. But if they say the Bible is true, you say they are not 'serious.' Then you destroy their careers. You have destroyed the careers of hundreds of brilliant scientists."

Eli continued. "The problem is, when you look through the lens of materialism, you see only material things. You filter out true love and you filter out true beauty. Worse, you filter out God.

"You search for meaning, but you are blind to it. There is no meaning in the material world, and the material world is all you can see. You cannot see the living God. So, you worship idols, false gods, and pretend they have power. Some worship the false god of money, they think about money all the time. They cannot get, and never will get, enough money. Others worship power, and when power is your god, the end always justifies whatever means you use to acquire it. You will torture, you will kill, you will invade Ukraine. Some worship the pleasures of the flesh, the pleasures of drink, of drugs, of sex, and by so doing, they are doomed.

"The false gods will never satisfy you. You always know, even when you won't admit it, even when you are consumed by money, power, or the pleasures of the flesh, you always know. You know death will take it all away, that whatever you have amassed or whatever you have achieved, or whatever you have dreamed, death will destroy it. There is no escape. You see no way out. You are angry. You are doomed, you are without hope. You have fallen for the lies of the Devil. When you see only the material world, you are doomed. You have filtered out God and you have filtered out the Devil. You cannot see the spiritual battle between God and the forces of evil."

John could feel Eli's excitement at sharing his thoughts. Ashley and Hajid seemed confused and a little uncomfortable, while Svetlana wore a scornful sneer and Mark appeared bored. Mary was back in the family room. Fourth was finishing his fourth martini.

Eli smiled, almost glowing as he spoke. "I see creation through a different lens. I see beauty everywhere. I see profound love, love beyond understanding. I see the love of God. I know with every breath I take that the Creator of the universe took human form and offered himself as a sacrifice, as the perfect sacrifice, not only for my sins, but for the sins of the entire world. I cannot understand such love. I cannot fathom its depth. Somehow, and I don't pretend to know why, God so loved the world that he gave his only Son, that whoever believes in him should not perish but have eternal life. [35] The love of God, the true light, shines through my lens with power.

"Yes I see suffering and evil, great evil, and I weep. Everywhere, I see lies of the Devil. I know this world, the material world, is doomed by the law of death. I walk, we all walk, through the valley of the shadow of death. But I know that my redeemer lives, and at the last he will stand upon the earth. [36] And I know, with absolute trust in the living God, that he will raise me up on the last day.

[37] I know in this world I will have suffering, but I know, I trust, I see through my lens with absolute clarity, that Jesus Christ has overcome the world. [38]

"We are born with a lens of sin, a lens that sees only the material world, a lens that wants only the material world. God invites us to rise above. The living God invites us to join him in paradise, to see Creation differently, to put on a new lens, which he will give us if we ask. This is what Jesus meant when he spoke of a new birth, of being reborn in the spirit. [39] For unless we are reborn of the spirit, unless we see the world through a new lens, we cannot see God. And if we do not see God, we are doomed, we have nothing. We cannot escape, we will not escape, the lies of the Devil.

"The living God stands at the door and knocks. He invites us to open our hearts, to see his creation through a lens of faith, a lens of trust in the living God."

Eli stopped talking. You could have heard a pin drop. John was overcome by emotion; Mary reached over and squeezed his hand. Matthew and Imani were smiling. Ashley and Tiffany looked puzzled, like their heads were full of questions. Svetlana seemed annoyed, as if she shouldn't have to listen to such nonsense from a plain, despicably dressed, and pathetically uneducated person. Mark wasn't sure whether to proceed with Plan A, Plan B, or Plan C-cret.

Just then, Hajid's demon returned. He was once again a tenured professor, all-knowledgeable and intellectually superior in every way. He was the world's expert on the material world, and he was going to defend it.

CHAPTER THIRTY-EIGHT

Hajid was angry. The spell of Imani's voice had worn off. No tenured professor should have to listen to such nonsense. Time to stand up for materiality and take down this stranger. Did this kid think he could fake out a world-renown Harvard professor on quantum physics? It was time to destroy him. *And maybe*, thought Hajid, *I will finally get some respect from my uneducated father-in- law.*

Hajid leaned forward. "I don't see your point. You avoided my question. What about the laws of physics?"

Eli was ready. "I did answer your question, but you can't see it. The beginning of atheism is the belief there is something other than God. But God is all there is."

Hajid almost sneered. "And the material world?" "The material world is made up of the thoughts of God. True science reveals it."

"And exactly what branch of science is that?' asked Hajid, raising his eyebrows. He knew what Eli was going to say, and his demon was ready to pounce.

"Quantum physics."

That was it. Hajid's demon roared. "Ha! Do you have a clue, the slightest clue, who you are talking to?" His voice boomed with indignation. "I teach quantum phys-ics! I lecture around the world! I write papers published in top journals! I am *the* go-to expert on quantum physics! You have no idea who you are talking to!"

Perhaps Hajid expected Eli to be intimidated. Perhaps Hajid expected Eli to recoil in apology. But it was not to be. Eli did not appear to be impressed, not in the slightest. Eli sat and smiled, while Hajid seethed. Obviously, this kid was not smart, at least not smart enough to realize who he was talking to or when to quit.

"Good," said Eli. "Then I can explain it to you. But let's be clear. Only God can open your eyes to see it. I can explain it, but I can't make you see it."

Hajid was stumped for words so he let Eli continue. "There is a scientific experiment that proves that the material world is not all there is."

Hajid shook his head. This kid was an idiot. "And exactly what experiment might that be?"

"Quantum entanglement," stated Eli simply. "Parti-cles can be connected outside space and time."

"We just haven't figured that out. Someday we will," Hajid countered.

"And when you do, you will be closer to God. There is no explanation in space and time for how particles can be instantly connected, with no delay whatsoever."

John interrupted. "What's this experiment?" "Physi-cists like Hajid can take two particles and 'entangle' them, connect them, make them identical," explained Eli. "Then they separate the particles, sometimes miles apart. When they change one, the other changes—instantly, with no delay whatsoever, not even the delay needed for light to travel between them. This violates Einstein's theory of relativity, which says nothing in the material world can travel faster than the speed of light. This experiment proves there are connections outside the material world, outside space and time. By itself, this experiment destroys materialism. It proves the material world is not and cannot be all there is. By itself, this experiment destroys Hajid's worldview, but he cannot see it."

Hajid's demon did not respond. This kid was not backing down.

"And that's just the beginning," continued Eli. "Another experiment is the one-slit/two-slit experiment, where particles can be in two different places simultaneously."

Now everyone was confused. It didn't make sense. It was John who asked, "That sounds impossible. Can you explain?"

"It's complicated, but particles can go through two holes at the same time. I know it doesn't make sense. But it's true. It's been proven. The one-slit/two-slit experiment proves the universe is mental, not physical. Deep down, beneath the illusion of materiality, the universe is made up of thoughts, not particles as we think of them."

Eli looked at John and then Hajid. He smiled with joy. "We live in the mind of God. That's what the Bible tells us. That's exactly what the Bible tells us. We are not far from God; God is within us. In him we live and move and have our being. [40] If you want to see the mind of God, open your eyes, you're in it."

Eli kept going. "The Gospel of John tells us 'In the beginning was the Word.' [41] The Greek word at the end of that sentence is 'logos,' which means 'divine thought.'

The Bible tells us God thought the universe into existence."

John interrupted. "Why? Why did God do that?" "God only knows," stated Eli. "Only God knows." He

smiled. "You can ask him when you meet him. And you and I and everyone else will meet God, we will have a sit- down, kind of an exit interview, with Jesus. But I think God created the universe for us because he loves us and wants a relationship with us. The Bible tells us so. The divine thoughts of God created the universe and beings with the capacity to love him. But love is not real if it is forced. God gives us free will, and we choose."

Eli took a deep breath before continuing, "But let's get back to physics, and the answer to Professor Akbas's question. His question was, 'How can the laws of physics be violated?' Hidden within that question is an assumption, the assumption that the laws of physics control the material world. But they don't. God controls the material world. He created it all."

Eli looked at Hajid. "You see, these experiments, and they have been repeated thousands of times, destroy your concept of reality. The one-slit/two-slit experiment proves the universe is mental, and at the most fundamental level, particles melt into mathematical beauty, into thoughts. If Einstein's theory of relativity is true, and nothing in space and time can go faster than the speed of light, then there are undeniable connections outside space and time. Space, matter, energy, and time are not all there is. There is something greater, something outside.

"All the evidence, when you look at it honestly, points to wonder. I think of the universe as a fishbowl. Yes, it is immense. Yes, it is incomprehensibly big. And yes, we cannot see outside it. Perhaps we are less able to understand what's outside than a fish in a bowl. The fish may see objects; we see nothing. For the fish, what lies outside the bowl is made up of the same stuff as what's in the bowl, space and time and matter and energy. But we have no clue what lies outside. We are inside the thoughts of God. We can only know about God that which he has chosen to reveal. God has given us the Bible, and all the saints and the prophets, to help us understand him, his power and his glory, his mercy, and his love for us."

Hajid's demon refused to yield. "You make it sound simple. If we live in the mind of God, if he thought the universe into existence, why didn't he think happier thoughts? Why is there suffering? If God is good and all powerful, why do we suffer?"

"Fair enough," Eli nodded. "For that, we need to talk more about what lies outside our fishbowl, this fishbowl we call a universe."

CHAPTER THIRTY-NINE

Imani had been listening. She didn't know anything about physics. But she did know about suffering. She had experienced enough of it. Because she was feeling more comfortable now and everyone had enjoyed her singing, she thought maybe no one would mind if she spoke up.

"Suffering brings us closer to God," Imani chimed in. "If we didn't suffer, we'd be spoiled brats."

"But why?" asked John. "I have problems with that answer. Why couldn't God have designed it some other way? Why couldn't it be we get closer to God just by walking on the beach or admiring a beautiful sunset?"

Imani wasn't sure how to answer, but Eli knew exactly what to say.

"I agree with both of you. Suffering does bring us closer to God. It is the best antidote for the Kool-Aid of the Enlightenment, the arrogance of those who deny God. And we shouldn't be surprised we suffer. Not if we read the Bible. God makes it clear—'in this world you will have suffering.' [42]

"But walking on the beach, by itself, doesn't do it. That's the poster for new-age spirituality. Nothing wrong with it, God wants us to enjoy his Creation. But by itself, the act is meaningless and empty. It's for people who say they want to have a relationship with God but are too lazy to do the work a relationship requires, like going to church, praying, and reading the Bible." Eli looked at Fourth. "It's for people who want to worship themselves."

Imani stifled a giggle. Fourth was taken aback. But he didn't respond, perhaps his brain was not at its sharpest setting. Of course, that's assuming there was a sharper setting in there somewhere, which was open to debate. Before Fourth could think of a reply, Eli continued.

"The Bible tells us we suffer, and God's Creation groans in agony under the law of death, because of the fall. Because of the disobedience of Adam and Eve. Because of our human pride, our human arrogance, that we can do it, make it, without God. That through sheer human intellect—what you mistakenly and arrogantly call 'science'—we can do it all. But how can the folly of human intelligence dare to compare with the majesty of God?"

Hajid cleared his throat. "But why? Why do we suffer? Answer that if you can. Explain misery. Explain the pain of billions. Why does your God, your God of so- called mercy, allow it?"

Eli leaned closer. He looked into Hajid's eyes; he was staring into Hajid's soul. "We are fallen beings, and we live in a fallen world. We can't understand it because it's outside our fishbowl. We are like goldfish speculating on what lies outside. God doesn't tell us because he knows we can't understand it, or we will refuse to understand it, or both."

Eli sat back. "That's it. The Book of Job makes it clear that, after the fall, God doesn't give us any other explanation. In the Book of Job, God lets the Devil inflict great suffering, seemingly unbearable suffering, on Job. Job's friends tell him to curse God and die. But Job doesn't. Job hangs in there, through the suffering, although he does complain and become angry with God. Then God speaks with power. Here is what God says:

"Where were you when I laid
the foundation of the earth?
Tell me, if you have understanding.
Who determined its measurements – surely you know!
Or who stretched the line upon it?" [43]

"You see, God is God. We want to be like God, we want to understand it all, but we are not God, we can't understand it all. We live in a beautiful universe, a universe of wondrous creation and design. But it's like a fishbowl. We can't see, we don't know, what is outside. We have no clue. We cannot see or begin to understand the magnificence of God. God knows the stars by name, and there are about as many stars as grains of sand on the whole earth. God knows each hair on your head. God thought the universe into existence. God is more magnificent than we can imagine.

"God doesn't tell us why we suffer. But we do know that suffering has a purpose. [44] And he came down to earth and suffered with us. He assures us that, if we want a relationship with him, if we seek him, he will comfort us. We need to have faith, and that means trust, real trust, trust in the living God."

Hajid sat silent. His demon had lost control, and the wonder of Eli's words began to seep in. Eli continued.

"And Imani is right. Suffering is like a megaphone. God uses it to call us out of our smug shells, to help us realize we can't do it on our own, to help us realize we need God. And if you realize you can't do it yourself, you are open to the Gospel.

"There are things we cannot understand. We cannot understand how God can be three persons in one, and we shouldn't expect to be able to understand it. But we know he is, the Bible tells us so, even in the beginning, in the Book of Genesis, where God states—'let *Us* make man in *Our* image.' [45] The original Hebrew is plural, plainly plural. Read it again, the words are clear. 'Let *Us* make man in *Our* image.' The Trinity is a concept we cannot begin to understand. It is outside our fishbowl."

Eli looked at Fourth. "There are many differences be-tween you and me, and between our worldviews. One is the purpose of life. You drink the Kool-Aid of the En-lightenment, you think the purpose of life is to maximize human happiness. I read the Bible, and I know the purpose of this life is to seek God so Jesus will take me with him to Paradise. That is the Gospel."

Eli looked at John, sensing his uncertainty.

"Help me…" John's voice tapered off into a plea, "help me understand. All my life I've heard the word 'Gospel,' the 'Gospel's message,' preaching 'the Gospel,' and so on. But I've always been confused. What is the heart of the Gospel? How does it work? They say 'Jesus saves.' What does that mean, and what must I do to be saved?"

CHAPTER FORTY

Imani had taught the Gospel; she had heard these questions before. She looked at Eli. "May I?" Eli gestured for her to go on.

Tiffany picked up her chair and moved closer. She wanted to listen. Something was missing from her life, something important. After eight years, the excitement of photo shoots and magazine covers had faded, and, perhaps without quite realizing it, she needed a new reason to get up in the morning. She was searching for something and perhaps, just perhaps, this was it.

Imani began. "My words may be simple, my theology crude. But, to me, the Gospel is not hard to understand." Imani looked at Matthew. He was beaming with pride.

With his silent blessing, she continued.

"'Gospel' means 'good news.' The good news is God allows us to trade, to exchange. We can exchange our brokenness for his righteousness.

"It's not hard. The first step is to admit you are broken. Admit you need God. And it has to be sincere, you can't fake it. You can't say, 'Hey God, I'm doing pretty good but if you wouldn't mind fixing up some parts of

"

my life that would be great.' You can't say 'Hey God, I'm really proud of myself for being smart, rich, good-looking, talented, or whatever, and I did drop that $5 bill in the collection plate, so please bless me and I'll be out of here.' You can't approach with pride. You have to admit you are broken.

"The second step is to realize what Jesus did for you on the cross. He paid the price for your sins. He came to die, to be a sacrifice for your sins, and my sins, and the sins of everyone, the sins of every person who has ever lived or ever will live. God is just, and justice demands a price, a price for our rebellion against the magnificence of God. Someone has to pay; someone always has to pay. So, God pays it himself, in his mercy toward us and in his love of us. It's the most famous verse in the Bible, John 3:16:

"For God so loved the world, that he gave his only Son, that whoever believes in him should not perish but have eternal life."

"So, step one is to admit you are broken, and step two is to realize Jesus can forgive sins. Now step three. Ask Jesus to trade. You go to Jesus and you ask to trade, to give him your brokenness and get a piece of his righteousness. That's it, one, two, three. The Gospel is easy. Maybe even easier. Paul tells us that if you declare with your mouth 'Jesus is Lord' and believe in your heart that God raised him from the dead, you will be saved." [46]

Imani looked around at her audience. "And I get it, it makes no sense, no sense whatsoever. You give Jesus brokenness, and he trades you for salvation, for eternal life, for a place with him in Paradise? But he does, and he will, *if* you have faith. You have to ask. It's like communion. People take their troubles to the rail and, through a miracle beyond human understanding, through the bread and the wine that become the body and blood of Jesus, take in his righteousness.

"Jesus changes lives. The Gospel—this great exchange—changes lives. It's not something you can earn. Doesn't matter how hard you try, or what you do, you can't get there on your own. You can't earn your way to salvation and a life with Jesus. You and I, we don't deserve it. But Jesus gives it freely, if you ask. He paid our price. That's the Gospel."

The room fell silent. Tiffany smiled, the light of understanding was dawning in her eyes.

Imani looked to Eli, seeking his approval. Had her simple words been enough?

"Perfect!" said Eli. "Bible verses show it's true." Imani grinned. She loved Bible verses.

Eli began. "Start with the words of the prophet Isaiah, written 700 years before Jesus was born:

"But he was pierced for our transgressions;
he was crushed for our iniquities;
upon him was the chastisement that brought us peace,
and with his wounds we are healed." [47]

"You see, Jesus wasn't a nice guy who got a raw deal. He came to die. He came to be the sacrificial lamb of God. As Jesus told us, 'Greater love has no one than this, that someone lay down his life for his friends.' [48] Paul explained it in his second letter to the church at Corinth:

For our sake he made him to be sin who knew no sin, so that in him we might become the righteousness of God. [49]

"It happened at the crucifixion," explained Eli. "Even if you were there, you would not have seen it happening. It happened in the dimension of the spirit. As he hung on the cross for hours, Jesus took on himself the wrath of God, the punishment for all sins of the human race. The Old Testament saints knew this was coming. Today, we look back on it. But both are by faith, it's always been by faith. God's grace is an amazing thing. We cannot earn grace, otherwise grace is no longer grace but a wage. It is a great gift, the greatest gift, and we do not deserve it.

"Theologians call this the substitutionary atonement. It's the heart of the gospel. While he hung on the cross, Jesus paid the price for your sins, he paid the ransom to free you to join him in paradise.

"And Imani is right. You have to ask Jesus to trade, to take your brokenness and give you his righteousness.

"And if you do, Jesus will save you, he will save you from Hell. And you will be changed, your life will be changed. Jesus will begin to adjust you. You will be justified by faith. You will begin to take on his righteousness.

Some of the things you used to like will no longer please you. You won't want to hang out with people who are mean or petty. You won't want to lie or steal. It will change your life. Jesus will begin to adjust you, make you fit to dwell with him in paradise. You see, Jesus changes people. Jesus changes lives. Only Jesus can do that."

Imani smiled in wonder. "Yes, that's it! That's the Gospel."

"You see," explained Eli, "we live in a fallen world. We were made for Eden, to spend eternity with God. But we fell, all of humanity fell, when Adam and Eve disobeyed God, when they ate the fruit of the Tree of the Knowledge of Good and Evil. To make it right, as only God could, Jesus came and died on a wooden cross, in a way on a tree of death. I suspect the two trees—the tree of the Fall and the 'tree' of redemption—were in the same spot, or about the same spot. It's now Jerusalem, and it's also where Abraham went to sacrifice his son Isaac, when an angel stayed his hand. [50] I can't prove it, and we'll never know, but I suspect there are many reasons why Jerusalem is the holy city of God."

There was a pause, a collective intake of breath, as Eli's words dissolved doubts and penetrated souls. Mary broke the spell by announcing that Christmas dinner was ready. The feast was about to begin.

CHAPTER FORTY-ONE

Mary was in charge of carving the turkey. She insisted the breasts be removed in one piece and then sliced and the dark meat arranged a certain way. She never let John do it, said it was not exactly in his skill set. The table was magnificent. Crystal decanters sparkled and platters piled high with delicious food sat amidst the elegant place settings.

John looked at the gifts of God and smiled. He felt contentment, the peace he had felt on the ride out of the city, the peace beyond understanding.

John snuck a glance at Eli. Something was going on, something profound. Eli's words had power. They were a door to a greater reality, a door he was now eager to step through. But who was this skinny young man of plain face? A young man who more than held his own in a discussion with Hajid on quantum physics and quoted the Bible with authority. John had no idea, but he asked him to say grace in any case.

Eli bowed his head. "Bless this food to our use and us to thy service. In Jesus's name, we pray."

***

Ashley was relieved grace was over quickly. After the long discussion in the family room, she had been expecting some drawn out speech.

Ashley was delighted by her father's recovery. She loved him and was immensely grateful for all the things he did for her and her family. She most certainly didn't want him dead. She had hoped to convince her father that science, the worldview that there is nothing other than material things and that human intellect can do it all and solve any problem, was the way, the *only* way, to save the world. But now, she wasn't sure. Now, she had doubts about her prior doubts—could true science point to God? Now she was skeptical of her prior skepticism—had she been blind to wonder?

But still, she thought, at that moment, her father should leave his money to science. That would undoubtedly do the most good.

But she was puzzled. Who was Eli? She looked at the stranger. Who was this tall Asian-American young man dressed in shabby clothes, with a man-bun and ragged stubble, sitting in their world-class dining room? He seemed kind, he seemed nice, but there was no visible evidence of education. Where did he learn quantum physics, and how could he speak with such power and conviction?

Ashley was not the only person who wanted to know.

Her mother asked before she could take a bite.

"Eli, please tell us about yourself. Where are you from and what do you do?"

Eli smiled kindly. "I was born in China. My mother was a missionary. We emigrated to Vancouver when I was small. We didn't have money, but I got a scholarship to a technical school near Boston."

Ashley couldn't help herself. "What school was that?" "MIT."

The color drained from Ashley's face. Inside her mind, the wheels were spinning. Then something clicked. She had to ask.

"And your last name is Wu?" "That's right."

"Is Eli short for Elijah?" "Yes."

"Are you the Elijah Wu who conquered the Putnam?" Elijah smiled.

Ashley pressed on, determined to get an answer. "You're the Elijah Wu who got a perfect score?"

Elijah smiled and nodded slowly. "Guilty."

Ashley shook her head in disbelief. Elijah Wu was a legend at MIT, almost a myth. He was the greatest mathematical genius the school had ever known, with talent beyond that of Ashley and every other professor at MIT. And here he was, sitting at her parents' table, eating Christmas dinner.

Maybe Eli wasn't so dumb.

CHAPTER FORTY-TWO

Mark couldn't wait to change the subject. After all, he was the one who picked the wines, the two wines of quality, that is. No telling what two-buck-chuck this Eli/Elijah brought. And since anything that had to do with MIT was by definition painful and boring, and he had suffered enough sitting through that tedious discussion in the family room, he interrupted. "What do you think, Dad? Which wine do you like best?"

That was their game, their traditional Christmas challenge. Usually, they each picked a bottle of wine and guests tasted them blind and voted. This year, Mark had chosen both of the wines, but that was okay, he didn't mind the extra work. They could still have fun arguing which was best. It would be good for his father. It was Christmas, and at Christmas, you stuck with tradition.

In front of John were three wine glasses, each with a small pour. John picked up the glass on the right. He held it to the light and admired the deep color. He swirled the wine, put his nose in the glass, and smelled its bouquet. He took a sip and let the wine wash around in his mouth. John didn't know it, but it was a Grand Cru Bordeaux his wine consultant had bought at auction for $4,000 a bottle. "The first wine is good," said John. "Very good. I like it."

John picked up the next glass and went through the same routine. This was a cabernet from a small estate in Napa Valley. It had been aged for over 20 years, and there were only a few bottles of that vintage left in the world.

"This is really good," said John. "Maybe better."

John picked up the final glass and noticed Mark smirking. *What was that look for?* he wondered.

Mark watched his father take a sip of the wine. There was no reaction. Then his father smelled it again and took another sip. Then a third sip.

"There is no question," said John. "This third wine is the best. I think this is the best wine I've ever had. Please fill this glass up. Now tell me, Mark, which wine did I pick?"

Mark was lost. He watched as person after person tasted the wines. All agreed the third wine was the best, everyone other than Matthew. When Matthew reached for the wine, Imani put out her hand and gently stopped him. But everyone else, not counting Rebecca, who was sneaking glances at her cell phone, was raving about the third wine. Mark tasted the three wines himself. It can't be, he thought. But it was. There was no way, but no denying, the third wine, Eli's wine, was the best. Mark heard his father again. "Tell me, Mark, which wine did I pick?"

Mark found it difficult to spit the words out. "That's the wine Eli brought."

There was laughter, cheers. Glasses clinked. Mark's face turned red; he had been humiliated. He watched as the decanter with Eli's wine was passed around. Everyone who was drinking wanted a full glass. And there was no mistaking. It was right in front of them, plainly written on the piece of paper Mark had taped on the decanter. *Eli's wine.*

CHAPTER FORTY-THREE

Wine tasting was always fun, but Mary was more curious than ever. "What's the Putnam?" she asked Ashley. "And how do you know Elijah?"

"I don't know Elijah," Ashley answered. "But he is a legend. Elijah was creating new mathematics and publishing papers his first year at MIT. Then he got a perfect score in the Putnam Mathematical Competition." Mary looked impressed, so Ashley explained further. "You see, the Putnam is impossible. Every year, thousands of the country's top math students take it. And most can't crack a single problem. The top score is 120.

It's a six-hour test, three hours in the morning and three hours in the afternoon."

Ashley realized this was a teaching opportunity. She looked at her daughter. "The median score is usually a zero. That's the same as saying that most of the people who have ever taken the test couldn't begin to crack a single problem, couldn't get the slightest bit of partial credit."

"I know what median means," said Rebecca, pulling a face.

"Of course, you do," said Ashley, smiling indulgently. "Of course you do. Anyway, as I was saying, the Putnam is an individual test, but the top scores at each institution are used to make it a team competition. The year Elijah took it, we won easily. We've now won, I mean MIT has won, seven of the last nine years. [51] When I was a senior, I scored in the top 15." Ashley paused to let that fact sink in.

"Anyway," she continued, "here is why Elijah is a legend. He got a perfect score the first time he took the test, when he was a sophomore. And he walked out early from each session, both in the morning and in the afternoon. They say he took only 30 minutes to answer all the questions in the afternoon.

"That's not all. In the middle of his sophomore year, shortly after getting a perfect score, Elijah disappeared. He dropped out of MIT. Nobody had a clue where he'd gone, nobody heard from him again. Until now."

Ashley turned to face Elijah.

"We still talk about you. Please tell me. How did you do it? And why did you leave?"

CHAPTER FORTY-FOUR

Then spoke Elijah.

"You ask two questions. One I cannot answer. One I will try. As to how I did it, I don't know. Answers came and I wrote them down. It puzzled me. I wanted to know, how *did* I do it? Where *did* the answers come from? To say they came from the brain is no answer. It only puts a name on the puzzle. What is the human brain? How does it work? I wanted to know.

"And not just my brain. I saw concert pianists play. How could their brains do that? My college roommate could destroy me in chess. He didn't have to look at the chess board. He could visualize the entire board and all possible moves in his head. How could his brain do that? "It was a puzzle. I kept thinking. Then I realized every person has a brain that is magnificent beyond belief, magnificent beyond understanding. How do brains remember? How do they process sights and sounds? How do they create maps in our heads that allow us to move around the world? How does our subconscious moderate breathing, heartbeat, and so on? Why are we self-aware? How do we know what we know? Every person has a brain that is magnificent beyond belief, magnificent beyond understanding."

Elijah paused and looked around the room.

"So, I started reading. I wanted to know. What is the human brain? Where did it come from? I learned the human brain has as many cells as stars in a galaxy, about one hundred billion. I learned those cells are connected in as many ways as stars in a thousand galaxies. I learned there are more processing units in each of our brains than in the most powerful computers on Earth. I learned our brains are 1,000 times more energy efficient than man- made systems.

"And then, and I don't remember why exactly, I came across an article that stopped me cold. It sent shivers through my soul. I learned that to solve some problems, our brains build hundreds of millions of structures, incredible structures with incredible geometries. [52] I couldn't stop thinking about how that could be and where that design came from. But it made sense. I thought maybe, *maybe,* that's why when you do something over and over, your brain gets better at it. Maybe your brain gets better at building those structures. And then I thought, where did that come from, how did the brain learn to do that? It was technology and design far beyond what I expected and far beyond any human design. I used to think the brain was some sort of gooey mush. I never thought about how it works."

Elijah took a breath, then a sip of wine.

"There I was at a place of technology. But the most fantastic technology in the universe was within me. And yet that place, and every other place and person in the world, had no answer to my question. How does the brain work?

"Then I had another question, a powerful question, a question that changed my life. Where did the human brain come from? Who designed it, who made it so wondrous, so powerful? I was at a place of technology, but MIT had no answer. In a way, the question was almost forbidden.

"Who designed the human brain?"

CHAPTER FORTY-FIVE

The dinner was undeniably delicious. The wine was ecstasy. Everyone was digging in and exclaiming in delight. The decanter with "Eli's wine" taped on it was passed around and passed around again. There was plenty to go around. Glasses were filled and filled again.

John was not surprised the Very High Reverend was enjoying the wine. The man seemed to have a natural talent for anything involving alcohol. At least he was sitting at the other end of the table.

John and Mary looked at each other. Mary took John's hand and squeezed it gently.

"It's all too wonderful," she said. "The kids are all here, and I've still got you. I was worried Matthew would drink; thank God Imani stopped him. He's sober, can you believe it? Tiffany seems wonderful, genuine, surprisingly down-to-earth, and Imani…" Mary paused, looked at Imani, and spoke softly. "Well, Imani is amazing. We are blessed, John, truly blessed."

John took in the wonder. Mary was right. You didn't argue with Mary on something like that. They *were* blessed. He didn't know why he was sitting there. He shouldn't be. He should be lying in pain, perhaps dead. But here he was. He was content and at peace. There was no explanation. He did not feel angry. He did not feel grumpy. And he did not feel judgmental. His demons, the demons of money and power, were gone, at least for the moment. Thanks be to God.

CHAPTER FORTY-SIX

Svetlana was enjoying the wine, and it was going down far too easily. But she was lonely, and she was angry. She kept thinking of her husband on their yacht with Buffy. Buffy—is that a name or a joke? And how could her husband leave her, after she had dedicated herself to science? The Italian she met in Switzerland was a boy toy; there was no meaning or joy in that relationship. Despite her fortune, her education, and her position, she was falling apart. Deep down, she was miserable, in pain, and the wine was making her realize just how much. Her world had crashed, and it seemed all hope had evaporated. To make matters worse—if they could get any worse—she had just about had her fill of listening to this unkempt college dropout. He might have won a math contest. Maybe he did or maybe he didn't, who cared? There were answers to the questions he was asking.

Scientific answers. Nobody else was speaking up. So, it looked like it was down to her.

"But there *is* an answer!" she said, a little too loudly. "There is *no* mystery! We've known the answer for 150 years! It's written in letters over three feet high on the school you went to! Six letters carved in stone—D-A-R-W-I-N! It's the greatest discovery in the history of science. It's the power of natural selection, the survival of the fittest. Life is constantly evolving and constantly improving. And now, after billions of years, we have the human brain. We know that from science."

Svetlana took another gulp of wine. Maybe she had been a little too forceful, maybe a tad insistent. But somebody had to do it! This kid had no idea who she was. It was about time he showed her, and science, some respect!

CHAPTER FORTY-SEVEN

Elijah took his time. But when he did speak, he did so softly and gently, and with kindness.

"Yes," he said. "Many say there's this magic power Darwin discovered. But, and this I swear on my Bible, no experiment agrees. You can't even design an experiment to test that. Because whatever we find, whatever technology is revealed, wonders like the human brain, or birds, turtles, and butterflies navigating by the Earth's magnetic field, people with fancy degrees recite the mantra of Darwin. 'It evolved,' they say, 'it evolved.'"

Elijah paused. Everyone was looking at him. This was not the time for a debate. It was Christmas, a time to celebrate the wonder of existence and the Word made flesh. He didn't want a debate; he hadn't come for a debate. He decided to keep it simple. He wasn't going to discuss the fossil record, that the thousands of intermediate creatures Darwin imagined were nowhere to be found. [53] He wasn't going to explain that *all* advanced computer simulations showed exactly the opposite of what Darwin predicted. Mutations, random errors in DNA coding, always lost information, they did not, and mathematically could not, create new information and build new technology. He decided to keep it simple, so they could get back to the dinner and the joy.

"The basic problem with Darwin," he explained, "is that his theory violates the most fundamental law in the universe. It violates the law of death. When God kicked Adam and Eve out of the Garden, death came into the world. And since death came into the world, his creation has been groaning in pain." [54]

Elijah nodded toward Hajid. "I think you know what I mean."

Hajid did not like being called out, but it was an impertinent remark and not hard to answer. "Well now. Perhaps you're talking about entropy or, more formally, the second law of thermodynamics. Everything rots, everything decays, and everything falls apart. Unless there is a mind to stop it, everything goes downhill."

"Yes," said Elijah. "I call it the law of death. We are under the shadow of death. Everything, all life, is dying. Things are changing, but it's down, not up.

"But technically," Elijah looked around the table, "it's not entropy that disproves Darwin. It's a cousin concept, the mathematics of the transfer of information. All organisms transfer information to the next generation. That's basic biology; DNA is copied. Simple math shows that when you copy complex information, things can only go downhill, unless something is acting on the system."

"But something *is* acting on the system," Hajid chipped in. "It's called natural selection. If the copying is no good, the organism dies."

"Yes," agreed Elijah. "But that is a weak effect, especially when you're talking about copying billions of letters of DNA code, like we humans carry. Humans have about 100 copying errors, mutations, in each generation. Natural selection is mostly a thumbs up/thumbs down process, either the organism reproduces or it doesn't. Common sense says that when you've got over 100 errors, and only one thumbs up/thumbs down vote, the errors are going to add up, accumulate, in every generation. And here's what's even worse for Darwin, what totally drives a stake through the heart of Darwin's delusion. Most of the errors, the vast majority, have little or no effect by themselves. Scientists say they sit below the 'selection threshold.' By themselves, they don't have a measurable effect on fitness. So, natural selection cannot see them, natural selection is blind to most of the mutations. But they add up, and their combined effects over the generations can be lethal, like rust accumulating on the body of a car.

"It's simple math. Our DNA is degrading, not improving. Human beings are going downhill. The magic Darwin imagined does not exist. Natural selection is not a force. It has no purpose, and it can't create. It can't design or build the human brain."

Elijah looked around the room, and his gaze settled on Ashley. It was time to answer Ashley's second question.

"I realized," continued Elijah, "there is only one answer. Only one answer to how I could solve problems. Only one answer to where the human brain came from. And that one answer is God. It was so obvious, yet strangely forbidden. I could not live my life in a lie. I was going to a school with a lie carved in stone in large letters on the main building. I couldn't do that, so I left MIT. I turned away from the trappings of academic superiority and the pursuit of worldly riches. I decided to follow Jesus.

"My life is now a search for the presence of God. I no longer study math or physics. I rejoice always, I pray without ceasing, and I give thanks in all circumstances, thanks to the living God." [55]

CHAPTER FORTY-EIGHT

R "But there's problems," she said. "If life goes ebecca had been quiet. But questions were dancing. downhill, how do we get new species? And how did life get started? My teacher said life began in a dirty pond. She said if you go back far enough you will find all life, all plants and animals, are descended from some first form of life. Isn't everything evolving for the better?"

Elijah turned to Mary. "I'm sorry. I didn't mean to start a debate."

"And it's a stupid debate!" interrupted Svetlana, again a little too loudly. "This is *settled* science! Darwin's the answer, the *only* answer!" Then, in a more subdued voice, "And would you please pass the wine, the decanter with the piece of paper taped on it?"

Mary watched Elijah pass the decanter to Svetlana. How could the decanter still be full? And what *was* that dark spot on Elijah's hand, the spot that was plainly visible when he reached out to pass the wine? Mary now realized that both of his hands had dark spots. But those were questions for later. Now it was her granddaughter's turn.

"It's okay," said Mary. "Please tell us. I have the same questions. Where do new species come from? How did life get started? If Darwin's wrong, what's right?"

Elijah smiled at her. "I'd be pleased to answer. Those are wonderful questions. Would you please pass the turkey and the wine? And could I have more of that cranberry sauce?"

Elijah refilled his wine glass, took a sip, and smiled. He put more turkey on his plate and took another bite. He put more cranberry sauce on his plate, and while he was at it, he added more stuffing, more green beans with almonds, more lima beans, more mashed potatoes, and, of course, more gravy. He was relaxed and confident. He radiated joy. There was something about him.

"This may sound strange," he began. "It is not what you have been told. But, on balance, it is more wonderful than strange." He swallowed a mouthful of potato.

"We live in a world of lies. The Adversary, the Devil, lied to Adam and Eve, and the lies of the Devil are strong today. Many have swallowed the lies of the Devil. Paul said it 2,000 years ago, and his words take aim across the centuries at the conceit of our world today: 'Claiming to be wise, they became fools.' [56] Mortal men have trusted in themselves rather than God. They worship their own intelligence rather than the immortal God."

Elijah paused and looked across the table at Svetlana, Ashley, and Hajid. "Some call their false religion science.

But true science is just a tool, just a way to figure things out."

Svetlana slammed her wine glass down. Fortunately, it was empty. She almost shouted. "You know nothing about science! Science will save us!"

Iron Mary did not appreciate that. She glared at Svetlana. "Rebecca has questions, and I've asked Elijah to answer. We *will* let Elijah speak."

Svetlana looked down. She said nothing. It was not her house, maybe she'd had too much wine. She wished she had never come. Her world was crashing, her marriage was lost. Her wealth was meaningless, and her worldview was under attack.

"Thank you." Elijah smiled at Mary. He looked at Svetlana with compassion. "I know these ideas upset you. When worldviews clash, minds recoil in pain. Science is a tool, a way to figure things out. Science can help us find the truth, but it is not the truth. It is not something to be worshipped." Elijah took a big bite of the cranberries in red wine.

Then Elijah smiled at Rebecca. "You ask two questions. Let's talk first about the origin of life. The Bible tells us God created all life, creatures of all kinds, in the first week, the week of Creation."

Elijah put a combination of the turkey with gravy and the sausage stuffing in his mouth. He chewed and swallowed. Then he turned to Hajid. "But false prophets preach the lies of the Devil and say life started by chance.

Harvard swallowed that lie, didn't it?" Elijah smiled and took a sip of his wine.

"What do you mean?" asked Mary.

By that time Elijah had a mouthful of lima beans, so everyone had to wait until he finished chewing.

"In 2007, Harvard launched an 'Origin of Life Initiative.' They tried to figure out a way life could have started without God. But the more they looked at it, the more science revealed the majesty of life and the huge chasm between life and non-life. Life is far, far too complex to have arisen by chance. You can't go from chemistry to biology. Would you agree, Hajid?"

Hajid's eyes glittered with anger; he sat upright and glared at Elijah. He spoke with his professor voice, in a tone that suggested only a fool would dare to disagree. "Well, we don't know! Some think life was seeded from outer space."

Elijah chuckled loudly. "Yes! That's how it works! Don't you see?"

Elijah looked around the table. "Don't you see? When one lie is exposed, the Devil comes up with a bigger lie. I'm sorry, but life from outer space is absurd, totally absurd. It doesn't even try to answer the key question— how could the complexity of life have begun without a designer?—and it creates more problems than it pretends to solve. You still haven't figured out how life began and you've added the absurd suggestion that fantastically complex molecules can survive in outer space. Don't you see?

That's standard practice for the Devil, a bigger lie to hide the first. Other examples are an imaginary multiverse machine that somehow creates new laws of physics, and this 'punctuated equilibrium' nonsense that new species burst into existence for no reason."

"Punctuated equilibrium is a theory of a famous Harvard scientist," protested Hajid.

"And it's nonsense," stated Elijah, "total and obvious nonsense, a lie of the Devil. And everyone honest knows it." Elijah sat up straight, his presence becoming more commanding every second. He took a bite of stuffing.

Mary leaned toward Elijah. She wanted to learn more. "But what about new species? How do we get new species?"

"God builds the technology in," announced Elijah. "One of the ways he does it, if you need a name, is with intergenic ORFs—'open reading frames' in the DNA, the coding created by God—of all species. The sections of DNA code that tell our bodies how to build different proteins—the sections we call 'genes'—have between them and around them other sections of DNA code with extra technology that builds new species when needed. People used to call all this extra DNA 'junk DNA,' they had no clue what it did. Now we know it is fantastically advanced technology that creates new species when needed. That's how Noah fit the animals on the ark. He only needed one pair of dogs, one pair of cats, and so on, one pair of each kind of animal."

"Oh, come on! That's preposterous!" cried Svetlana. She was in the middle of refilling her wine glass and frowning at the decanter with Eli's name on it. *Maybe she's thinking what I'm thinking,* thought Mary. *Why is so much of it still left?*

"It's the truth," said Elijah plainly. "It's been proven. God builds the technology in. Scientists take normal fish and put them in a dark tank, with no light, and acidic water like in a cave. Complex systems on the fish change. Their skin changes, gills grow larger, and so on. Don't you see? There are other examples. The technology to adapt is built in, as only God could do.

"Here's another. People in tropical regions have dark skin. God built that in to help them thrive in the hot sun. It's a complex system; dozens of genes regulate skin color. Darwin's theory of racial superiority is a lie of the Devil."

"Yes!" exclaimed Imani. She glared hotly at Svetlana. "Tell the truth! Admit it! Because of Darwin, Europeans slaughtered people of dark skin and hunted them for sport in some places. Because of Darwin, Hitler tried to kill all Jews. Because of Darwin, your ancestors put my ancestors in zoos."

Mary put her hand to her mouth in shock. "I didn't know that!"

"It's true," declared Imani in rage. "Criminal. And not that long ago. All because of Darwin. In 1906, the Bronx Zoo put a black man from the Congo in a monkey cage. And in other places, not just the Bronx. Black people were put in zoos in Chicago, Paris, London, and Milan. All because of Darwin, and his phony theory that white people are superior. I never thought of it the way Elijah puts it, but he's right. The Darwinian delusion that species magically improve not only denies God but also leads to hatred and bigotry. Hatred and bigotry that demeans both black and white. Millions of people have been killed because of Darwin's delusion that some people are superior to others."

The debate was getting heated. That didn't stop Rebecca. "What about the 'Tree of Life'? I've seen pictures. Doesn't everybody agree life used to be simple and then got complex? In the pictures, there's some tiny creature at the bottom, at the bottom of the trunk of a tree, and then the tree grows up with many branches and life 'evolves' into plants and bugs and fish and animals. Isn't that true?"

Elijah smiled kindly. "No. It's not. The picture of the Tree of Life is false. You might say science has cut it down. When they compare pieces of DNA—genes—across widely different creatures, they get different patterns, different connections, depending on which family of genes they look at." [57]

Rebecca frowned in confusion and Elijah laughed softly. "I can make it simpler. Here's my favorite. Have you heard of loggerhead turtles, large turtles that lay their eggs in the sand on the beach?"

"Yes!" she exclaimed. "We went to Florida last year. I didn't see turtles, not that kind of turtle, but I did see places where people had put sticks in the sand with bright orange tape, to keep people from stepping on turtle eggs in the sand."

"That's right," Elijah nodded encouragingly. "Did you know those turtles—and they're big, some weigh hundreds of pounds—swim thousands of miles in the open ocean, and come back years later to the same beach, the exact same beach, where they were born, to lay their eggs?"

Rebecca had a look of disbelief. "No." "It's true." "How?"

Elijah grinned. "Yes, how? That's the question. They do it—and I'm not making this up, it's been published around the world—using the earth's magnetic field, plus other technology that's built in."

There was silence. Elijah continued. "It's incredible, but those turtles have built-in technology, the natural ability, to measure the earth's magnetic field exactly, measure both its direction and its strength. They can sense the invisible magnetic lines of the earth. This is advanced technology—people have only figured it out recently—and it's complicated, because the earth's magnetic field keeps changing, like a dying fire that slowly, very slowly, is getting weaker."

Rebecca's eyes were never so wide. "Wow! *That's so cool!*"

"Yes," Elijah smiled. "It's more than cool. It's wonderous. It is wonderous technology only God could design, and only God could put in a turtle. And here's what proves the Tree of Life picture is not true. This technology is not just in turtles. God put this technology, this built-in ability to sense and follow the earth's magnetic field, into some birds and fish, like arctic terns that fly non-stop for thousands of miles and salmon that return to the stream where they were born."

Elijah saw Rebecca smile. "Are you ready for more?" "*Yes*!" Rebecca blurted out. "I love this!"

"God put this technology into some insects." "Bugs? No way!"

"Yep," said Elijah emphatically, "bugs." He gave Rebecca a mischievous grin. "But beautiful bugs. Monarch butterflies, gorgeous butterflies with orange and black patterns on their wings, can sense the earth's magnetic field. Some are born in Mexico, but migrate to the United States and even Canada, and return, four generations later, to the same tree."

John interrupted. "What? That can't be true. Four generations later?"

"Four generations later. Some return to the same tree where their great-grandfather emerged from a cocoon."

"How can that be?" asked Mary, her eyes wide with wonder.

Elijah smiled. "God only knows. He designed it." "That's *really* cool!" said Rebecca.

"But how?" asked John. "How could a butterfly know where its great-grandfather emerged from a cocoon?"

"Only God knows. It may be beyond human understanding. I doubt any person in the world has a clue how it is even possible. It is like the human brain. We are face-to-face with technology so advanced it is beyond our ability to understand. Maybe someday we will figure it out, maybe someday we will learn how God does it, how God passes knowledge from one generation to another. Today, we don't know. But it's true.

"You see, the Tree of Life is a myth, a false story, invented to allow human beings, particularly white Europeans, to look down on other creatures and even other people. God created creatures of different 'kinds,' creatures like cats, dogs, horses, and elephants. Then, in each 'kind,' he built in technology that allows that 'kind' of creature to change and adapt. That's how you get new species. At least, it's one of the ways you get new species. We're still learning. It's fantastic technology, technology only God could design, technology orders of magnitude beyond human technology, and we are just beginning to realize it is there.

"It's plainly in the Bible. The first chapter of Genesis tells us nine times that God created creatures 'according to their kinds.' The cat 'kind' includes lions and tigers; they can interbreed. The horse 'kind' includes donkeys and zebras and so forth. So, instead of one tree, the diversity of life is more like a forest, and each 'kind,' each basic type, is like a tree that has branches—different species. And God created it all."

Elijah took another sip of wine. "The key thing," he said, "is that it's down, not up. Simple math shows you can't build the fantastic technology of life by Darwin's simple process of keeping the best mistakes. Accidental mutations can't magically build technology. Evolution is down, not up. Accidental errors in the copying of our DNA are building up, like rust on a car, with more each generation. And not just in human DNA. All species are headed down."

Now nobody was challenging Elijah. Perhaps it was too strange. Elijah turned to Rebecca.

"Here's another wonder, a wonder that shows it's down, not up. There were giant animals on the earth, giants that have disappeared, giants that are now extinct. They've found places in Europe where people hunted elephants for food, but massive elephants, elephants three times the size of today's Asian elephants, elephants as big as 13 tons." [58]

"Where in Europe?" asked John. "And who did that?"

"People who lived in what is now Germany," said Elijah, "and perhaps other places, thousands of years ago. Some call them Neanderthals and imagine them as brutes, intermediate forms that magically 'evolved' into us superior modern humans. But that's chronological elitism, a cousin of racism, a horrible form of snobbery." Elijah looked towards Rebecca. "There were turtles 16 feet long that weighed over two tons. Sharks 80 feet long. Fifty-foot snakes and birds with 20-foot wingspans. Crocodiles 40 feet long and rhinoceros 18 feet high weighing 22 tons. Rats six feet long. And hundreds of others." [59]

Mary had one more question. "If all life is going downhill—if that is what science tells us—when did it start? How long ago was Creation?"

Elijah paused. He looked around the room as he nibbled on a green bean. Everybody waited. Elijah swallowed and smiled. "Great question," he told her. "Here's what I think, and this comes directly from the oldest texts of the Bible, directly from the word of the living God. About 7,500 years ago. Deep time, billions of years, is another lie of the Devil." [60]

CHAPTER FORTY-NINE

Elijah had thrown down the gauntlet. Worldviews collided; the fight was on. Everybody had questions. Some were angry, some just curious. Elijah remained calm, answering each question softly with eloquence.

"How can that be?" asked John. "What about Neanderthals? Didn't they live and die tens of thousands of years ago?"

"Yes and no," revealed Elijah. "Neanderthals were fully human; some think they had more DNA than we do. They had art and music and buried their dead. We carry their DNA within us. They lived mostly in Europe about 5,000 years ago, during the Ice Age. The climate was brutal, so many lived in caves and some suffered from vitamin deficiencies."

No one spoke. It was too strange.

"We are *not* smarter than people who lived thousands of years ago. That's chronological elitism, prejudice against people who aren't here to defend themselves. Yes, we have better technology and spend years in schools. Yes, we have knowledge and tools they could not dream of. But as for basic smarts, street smarts, we are not smarter.

That's chronological elitism, this conceit we are somehow smarter than people who came before.

"Consider Otzi the Iceman, who died 5,000 years ago from an arrow to the shoulder and was found frozen in the mountains between Austria and Italy. He wore waterproof shoes crafted in four layers with bearskin soles, deer hide uppers, cow leather shoelaces, and tree bark socks. He wore a coat of animal hides stitched together with animal fibers. He had a 'first aid kit' with over a dozen plants known to help heal wounds, cure disease, and fight parasites, and a fire kit with flint and tree fungus.

"Consider the Pyramids. The greatest structures ever built, and they were built 4,500 years ago. They are precisely aligned with the stars and seasons in many ways. The Great Pyramid has a central ceremonial room and an ascension shaft, cut and angled precisely through seventeen layers of huge stone blocks, that pointed directly to the belt of the constellation Orion. [61] The Egyptians believed a pharaoh became a star in Orion after he died. There are three pyramids because there are three stars in Orion's belt, and they are set beside the Nile River just as the stars in Orion's belt are set next to the river in the sky of our Milky Way galaxy. We are guilty of chronological elitism. The ancient Egyptians were not dumb; the ancient Greeks were in awe of them.

"What has survived from ancient cultures is the stones, and the stones speak to us today. The stones of the pyramids are cut precisely. Even today, 4,500 years later, you cannot put a knife blade between blocks that weigh tons. We couldn't build the pyramids today. In Peru, there are twenty-ton stones, curved stones, cut so precisely you cannot put a knife between them. In an ancient quarry in Jordan, there is a block of cut stone that weighs 1,500 tons, and there are 800-ton blocks beneath Roman temples. There are massive stone figures on Easter Island in the middle of the Pacific Ocean; no one knows how they got there or how they were taken up the cliffs. Stonehenge also illuminates ancient glory. It's down, not up. We should not be chronological snobs, and imagine ourselves superior."

Matthew had been quiet, but now he was puzzled. "What about the Ice Age? Weren't there lots of them?"

"No," said Elijah, and he smiled again. "There was only one Ice Age, and there is only one explanation for it."

Elijah turned to Imani. "And you know what caused it."

"Oh, come on," scoffed Hajid. "Please. This is nonsense. There were many ice ages and they were caused by slight changes in the Earth's orbit. It's been published."

Elijah grinned. "A lot of nonsense has been published. But the sun is strong today and was then, and slight imagined changes in orbit can't create ice sheets two miles high." Elijah turned back to Imani. "But *you* know."

Imani was quiet, so Elijah gave her a clue. "The greatest geological event in history."

Imani lit up. "Of course! The Flood, the Flood of Noah. Chapter 7 of the Book of Genesis. 'The fountains of the great deep burst forth,' and 'the windows of the Heavens' were opened, and the Earth was covered with water, even the highest mountains!"

"That's it!" chuckled Elijah. "Talk about climate change, right?"

"But when?" asked John. "This is strange, and it's wonderful. When? When was the Flood?"

Elijah looked around. Rebecca's eyes were glued to him. She was devouring every word. Ashley, Hajid, and Svetlana sat in scornful slouches; evidence was not going to alter their worldview. Fourth and Mark were bored and took the opportunity to refill their plates and wine glasses. Everyone else—Mary, John, Imani, Matthew, and Tiffany—was paying attention but stunned by the clash of worldviews.

"I love that question!" said Elijah. "When? I've studied it. It turns out, and this is not widely known, that there are two different versions of the Book of Genesis, which give very different answers as to when the Flood occurred and when Adam and Eve were created. Apart from these differences, the two versions are pretty much the same. It's clear one version or the other was deliberately altered as to the birth dates of the patriarchs, from Adam and Eve down through Noah to Abraham. I believe the oldest Bible texts, from a Greek translation almost 300 years before Jesus, are correct, and, according to those ancient texts and important non-biblical sources, the Flood was about 5,300 years ago." [62]

Elijah saw glazed looks. "If you want, I can come back to that later and explain how the Flood occurring 5,300 years ago—about 3,300 years before Jesus—fits with human history and non-biblical sources. But more important is how the Flood caused the Ice Age, and why a spectacular flood, the Flood of Noah, is the only way you get an Ice Age."

Silence. Elijah cleared his throat.

"The Flood was when God changed the laws of physics."

John sputtered, half in surprise and half in laughter. "But some people tell you the laws of physics cannot be violated!" He looked challengingly at Hajid.

Hajid frowned. He didn't take kindly to that remark, but he kept his mouth shut and said nothing.

"Yes," answered Elijah, "some say that, but we find patterns in rocks that cannot exist under today's laws of physics. It's technical, but the evidence, for those with eyes to see, is amazing.

"God changed the laws of physics, at least for a while, and underground oceans around the world exploded with hot lava and hot water. There were volcanos much larger and much more powerful than anything we see today. The Earth's crust cracked, and the pieces, the continental plates, hit each other like freight trains. Before the Flood there were no high mountains, not like the miles-high mountains we see today, so for a while the Flood water covered the entire Earth. Today's mountain ranges were created from the continental plates hitting each other. Even today, the Himalayas are rising."

Elijah paused. The room remained hushed.

He spoke softly. "It was beyond weather today. Skies around the world darkened with volcanic ash. Fantastic waves, tsunamis, the size and power of which we can hardly imagine, rolled across the Earth, and the Earth shook for months. When it was over, and the waters drained into today's oceans, the skies were dark and the water was warm. That's how you get an Ice Age.

"You see, the winters were cold because the ash from the volcanoes blocked the sun, but the oceans were warm, warm for perhaps centuries, from the heat of the volcanoes. The warm oceans gave off moisture, and snow fell for centuries. The snow turned to ice. In a century or two, there were ice sheets miles high in some places."

Elijah smiled kindly at Hajid. "No other explanation makes sense. The suggestion that slight perturbations in the Earth's orbit could trigger massive climate change is absurd, a lie of the Devil.

"And by the way," he continued, "the Flood of Noah also explains where fossil fuels come from. Some think that before the Flood there was more oxygen in the atmosphere. Before the Flood, before the fountains of the great deep burst forth, there was mostly land, and it was covered with thick and wonderous life. The tsunamis buried that life and created the pockets of oil and gas and coal we find today."

Rebecca almost shouted. "Wait! What about dinosaurs? They lived and died millions of years ago. Everybody knows that!"

Elijah smiled and looked at her. "Dinosaurs are cool."

"They're *awesome!*" said Rebecca. "I read every book I can."

Elijah leaned forward and spoke in a near whisper, as if he and Rebecca were sharing a secret. "What's your favorite?"

"Well, *actually*, they're *all* super cool," started Rebecca excitedly. "T-rex was a monster, 20 feet high, the scariest ever. And the Brontosaurus—or is it an Apatosaurus, I wish the books would make up their mind—was huge, some say 100 feet long. The Stegosaurus, you know with the bone plates that stuck up in triangles going down its back—so cool! And the triceratops, with the three horns, wow! I read it would beat T-rex in a fight, knock it over and crush it, and…"

Mark couldn't hold it in. He hadn't recovered from getting bested in the wine tasting. This was too much. Perhaps it had something to do with the wine, but he couldn't stand it any longer, he just couldn't.

He interrupted loudly. "Oh, come on! Seriously? I can't believe we're having this conversation. Rebecca nailed you. Dinosaurs prove the Earth is old. You sit here, in your t-shirt with a rip in the collar, and lecture us? I don't care if you won some math contest. This is nonsense."

Tension filled the room.

But Elijah still smiled. "Yes," he said, "when world-views collide, we recoil in pain. I know this is strange, but it is more wondrous than strange. There is more I could tell you if you have ears to hear, and eyes to see."

Mother Mary was not pleased. "I want to hear this!" She turned to Elijah in remorse. "I apologize for my son's rudeness." She glared at Mark, as only a mother can. "We *will* let Elijah speak."

Mark looked down, swirled the wine in his glass, and said nothing. He was angry, he wanted to shout out again, but he said nothing. He was not about to take on his mother, Iron Mary, especially in front of his father. And then he realized, perhaps too late, that his outburst was not good for Plan C-cret, not going to help convince his parents to let him handle their money. He looked up, put on a smile, and tried, as best he could, to seem sorry.

"Please forgive me. I don't know what got into me. I guess I'm drinking too much of this wonderful wine you brought. Please continue."

The air grew lighter, and Elijah smiled forgivingly at Mark. It was a smile that radiated kindness.

"I've been to Utah," explained Elijah. "I've seen it. There are natural rock bridges, carved by water from sol-id sandstone.

"Under one of the bridges is a brontosaurus. It was chipped away, carved *into* the stone, by the Anasazi Indians. Centuries ago, they saw a brontosaurus, and they carved it *into* the stone.

"I've been to Cambodia, to a temple called Ta Prohm, near Angkor Wat. There, carved on a stone column amongst a multitude of creatures, is a stegosaurus wearing what looks like a muzzle. Maybe it had been captured.

"There are other places. I've been to Carlisle Cathedral in northern England, 300 miles north and slightly west of London. Engraved in brass, over the tomb of a Bishop buried in 1496, are two dinosaurs, with their necks intertwined. One has spikes on its tail, of a type known by scientists. It's a depiction of a fight some people saw when two dinosaurs—two dragons—fought for over an hour."

Looks of disbelief reflected in every face. Elijah continued.

"It's written in ancient books, books we take as true except when they talk about dinosaurs. Dinosaurs used to be called dragons. Every ancient culture has stories of dragons. Alexander the Great encountered a dragon in India. Two Roman historians, centuries apart, wrote that India had the largest dragons. Marco Polo saw a dragon and wrote about it.

"Two centuries before Jesus, a Roman army was in North Africa attacking the city of Carthage. A dragon came up, behind the army, and the Roman general ordered his men to kill it. Many died, as the Roman swords could not pierce the dragon's skin. The dragon was killed using large stones from siege machines. They sent the skin to Rome. It was 120 feet long and on display for 100 years." [63]

Elijah looked at Rebecca, who was smiling widely. "And here's something really cool," he told her. "When people dig up dinosaur bones, they sometimes smell death, smell the faint odor of rotting flesh. They have found dinosaur skin, dinosaur blood. They have found flesh—soft tissue—from all kinds of fossils, all around the world.

"You see, every culture, every ancient tradition, has stories of dragons. And here's what's amazing but undeniably true. There are also hundreds and hundreds of ancient traditions telling of a worldwide flood. These traditions have many details in common! And, believe it or not, many of those details match the Bible! The ancient Hawaiians knew the name of Noah. There were eight people on Noah's Ark, and the ancient Chinese symbol for a large boat was a boat with eight people.

"There's a place in Montana. It's called Hell Creek, and for good reason. Something hellish happened there. There's a layer of rock, sedimentary rock laid down by water and debris from a tsunami during the Flood of Noah. Massive T-rexes—one had a skull that weighs 600 pounds—were immediately covered by mud, and are well preserved. The tissue is still soft, and one can even be identified as female. But here's why Hell Creek must have been hellish. Buried, with these dinosaurs, these huge dinosaurs, are massive sharks, at least six different types of sharks. And all the T-rexes and sharks and other creatures were buried, quickly covered in mud and buried together. Can you imagine? A wave so powerful that huge dinosaurs and large sharks are buried together!"

Elijah scanned the room. People were trying to process. Rebecca was smiling and eating it up.

"The Bible describes two dinosaurs, two extinct animals, in detail. It's at the end of the book of Job, which may be the oldest book in the Bible, it may have been written 2,000 years before Jesus. Job is suffering and angry at God, and God says look, I'm God, look at the majesty of my Creation, you're going to have to trust me. And God describes with pride two wondrous creatures he created, two massive dinosaurs."

Elijah looked at Rebecca. "One is a Brontosaurus, a creature with a huge tail sticking out straight, like a tree going sideways. The other is a fearsome beast with thick skin that swords and arrows cannot penetrate—perhaps that same beast that came up behind the Roman army at Carthage."

Elijah looked around. "I could go on, and on, and on. The evidence is all there, evidence of fantastic wonder, evidence that shows the Bible is true, from beginning to end, exactly as Jesus tells us it is. The qguestion isn't whether there's evidence. The question is whether you have eyes to see it."

Elijah paused. No one spoke. Silence reigned over the room.

"There's a well-known book, written by a Jewish historian named Josephus, born four years after Jesus died. Josephus fought the Romans, and then, through a twist of fate that only God could conceive, became the official historian for the Roman Empire. Josephus wrote a great book on ancient history, and he put it together using the combined knowledge of many older books in the Roman Empire. He lists by name four well-known, even earlier historians, who wrote that the Flood of Noah was a real event. He lists eleven earlier historians who wrote that the first humans, the ancients, lived for 1,000 years. Josephus wrote that both of these incredible statements—that the Flood was a real event, and, even more shocking at the time considering how short human life was during the Roman Empire, that the ancients lived for a thousand years—were in *all* the ancient history books.

"If you look closely, you will see the truth, and know for yourself that everything written in the Bible is true. But few have eyes to see. The lies of the Devil are powerful, and they are spread by people who, professing to be wise, have become fools."

Elijah took a breath to wrap up. "If you have eyes to see, go to the Grand Canyon. See for yourself, the rocks don't lie. See the straight lines between the layers. Look all around the canyon, and see that the lines are straight. You are looking at sedimentary rocks laid down by different stages of the Flood of Noah. Some of these layers of rock stretch underground for thousands of miles. The lines are straight. There is no erosion between the layers, absolutely no evidence of any erosion between layers supposedly separated by millions of years. There are places where they tell you the layers are separated by hundreds of millions of years. It's nonsense. There are fossils, mostly tree trunks, that are partly in one layer and partly in another.

"Go to the Grand Canyon. Go and see, and know that the Lord your God lives. And be humbled. For there is a beauty beyond anything you can see, and a reality beyond anything you can sense."

Again silence. It was wondrous.

Elijah smiled kindly at Svetlana. "It's not my job to convince you. I have my belief, and you have your unbelief. My belief is grounded in science, in history, in the stones of archeology, and the mathematics of life. I know you mean well, but, to me, your unbelief is a symptom of human arrogance."

CHAPTER FIFTY

John sat stunned. It was too strange, too wonderful. Then things changed. He entered a new reality, the dimension of the spirit. The dining room melted, and, before his eyes, Elijah floated above.

Demons rose to fight him. Each demon was a lie of the Devil. Out of Svetlana rose a great demon. It was the lie that life is ordinary. On its head was the mark of Darwin. It was pure darkness, with eyes of red fire. Its mouth was dripping blood from the souls it had devoured. From within it came the screams of a hundred million souls crushed by senseless wars, and the cries of a hundred million unborn babies.

John watched a demon rise out of Hajid. It was the lie the universe is ordinary. This demon had two heads. One was the lie that anything could exist without God. The other lie denied the majesty of God's creation.

A demon rose out of Mark. It was the lie of the false god. "Worship me," it hissed, "bow down and worship me." It was the lie of money, power, and pleasures of the flesh. It was fat from the flesh of a billion souls.

A fourth and final demon rose out of Fourth. It, too, was a monster of flaming evil. It was the lie of the so-called Enlightenment, the lie that life is about happiness and God has no place, the lie that human beings can rebuild Eden without God. In it were tens of thousands of lies about the nature of God, thousands of divisions in the one holy church of God.

The four great lies circled Elijah. Their eyes burned as of fire. They told Elijah they would devour him. They wore the smell of decaying flesh. Then there appeared, next to each demon, a pile of spears, a pile of false statements and beliefs, a pile of lies, and each was dripping with human blood. Each demon picked up a lie and faced Elijah.

John watched in awe as Elijah begin to glow. The armor of God was upon him. It was pure light, the armor of impenetrable truth. Elijah held a cross in his hand. The cross glowed against the darkness.

And then, in the dimension of the spirit, the battle began. One by one, the demons, the great lies, hurled powerful bolts of darkness. Each spear was a false statement or belief, a lie that had pierced the souls of millions. Elijah spun, faster than the eye could see, to deflect each. His armor shone. His cross beamed brighter.

The battle raged but Elijah did not fail. The light shone in the darkness, and the darkness did not overcome it. [64] The one true light, the light of God, grew ever brighter. The light overcame the darkness, and the great lies stood naked before it. Like phantoms, like shadows, they melted before the light.

The light was victorious. A multitude of angels began to sing. "Glory to God in the highest, and on Earth peace among those with whom he is pleased!" [65]

Slowly, slowly, the vision faded. It had been overwhelming, and John sat dazed. He blinked and shook his head. Now he was back in the dining room, and the house was playing "Go Tell It on the Mountain" at full volume.

Then, bizarrely, Elijah shouted. "Conga line!"

CHAPTER FIFTY-ONE

Nobody was expecting that, but they followed anyway. Elijah led the way, and the line formed behind him. Imani grabbed his waist, and Matthew took hers. Fourth was fourth. Tiffany was quick to join, with Mark just as quick behind her. He could not afford to look like a jerk again. Rebecca squealed with delight, and, an instant and a half later, Ashley and Hajid followed on behind her. John and Mary smiled at each other and took places at the end of the line. Only Svetlana did not move.

Elijah led them around the dining room, twice, and into the family room. The music grew louder:

"Go, tell it on the mountain
Over the hills and everywhere
Go, tell it on the mountain
That Jesus Christ is born."

Elijah showed them how it was done. "Go tell it on the moun*tain*," hips out, STOP! "Over the hills and ev*erywhere*," hips out, STOP! "Go tell it on the moun*tain*," hips out, STOP! "That Jesus Christ is *born*," raise hands and shout "Hallelujah!" The parade was unstoppable.

Maybe it was the wine. Maybe it was the Holy Spirit. They were possessed, and more so every minute. Legs kicked, hips thrusted, arms waved, and voices shouted as they danced into the family room and around.

When the song faded, they stopped to catch their breath. Laughter mingled in the air and the music changed. "Will the Circle Be Unbroken" began to play loudly, a country version with drums and a fast tempo. Elijah took John's hand, and they formed a circle. Rebecca stepped into the center and danced as everyone clapped to the rhythm. One by one, they took turns dancing in the center. Fourth resembled a bowl of shaking jelly. Tiffany sure knew how to move. Mark was cool, and Ashley managed not to embarrass herself. Hajid revealed his inner nerd, and, although John couldn't move too well, he danced because he was alive.

No one noticed as Svetlana put on her coat and walked out the back door. No one noticed as Elijah backed out of the circle, picked up his Bible, and walked to the front door. He put on his boots and coat, opened the door, and closed it softly behind him.

When the wind blows, and the sun goes down, it gets cold. Elijah walked into the snow. He closed his eyes and stopped. There exists a feeling that comes when one is alone in the cold and the dark and he felt it, he felt it in a way that cannot be described. He shivered in the cold dark of a moonless night—alone, a human being face-to-face with Existence. There were no words; there was only wonder.

He opened his eyes and looked up. Orion was rising. He said a short prayer, the Jesus prayer, the prayer that was always on his lips.

"Lord Jesus Christ, Son of God, have mercy on me, a sinner."

And then he was gone.

CHAPTER FIFTY-TWO

John was tired, and sleep came merciful and heavy. But then it was too heavy; then it was a nightmare. John was in Hell. Before him was Satan, aka Lucifer, the Adversary, the greatest of the fallen angels, the prince of demons, the ruler of evil spirits, the Devil himself. Around Satan were hundreds of fallen angels, and around them untold numbers of monstrous and misshapen creatures, seemingly going on forever, powerful beings of terrifying evil.

Satan shone with hideous beauty. He was crafted in gold and covered with precious stones. [66] He had been created by God to be the perfect angel, the signet of perfection, full of wisdom, and perfect in beauty. [67] He had been in Eden, the Garden of God. He had been an anointed guardian cherub [68] and had walked with God. But unrighteousness was found in him, for his heart was proud because of his beauty. [69] He wanted to be God. Then God cast him out of Heaven, and with him fell one-third of God's angels. He was a created being, bound by time, but with powers beyond mortals. He could not see the future and he could not see into Heaven, but he could see, if he cared to look, the state of all human beings and their world.

Satan had gathered the troops. There was Hell to pay. Satan thundered, and his words shook all of Hell. "Who is Elijah?!"

There was no response. And the demons clamored and shuddered.

Beelzebub, aka Baalzebûb, the fallen angel worshipped by the Philistines as their God, a powerful demon, a master of demonic possession of human souls, lieutenant to Satan himself, spoke. His voice was deep and ominous.

"Some say he is Elijah of old, the prophet from 2,850 years ago, the prophet who did not die but was taken up to Heaven in a whirlwind. My nemesis, who challenged and defeated 450 of my followers on Mount Carmel, by calling on God to send down fire, and had them put to death. [70] The prophet who mocked me, who said I was relieving myself when I could not rain down fire. [71] The Elijah who fled from my darling Queen Jezebel, but whom God renewed." [72]

Satan cut him off. "I know you burn for the Elijah of old. But is this the same?"

The mighty Beelzebub stared back. Then bent his head to his master and spoke softer. "I do not know."

Then spoke the Destroyer, the angel of the bottomless pit, perhaps the greatest of Satan's lieutenants. His name in Hebrew is Abaddon, and in Greek he is called Apollyon. [73] He has an army of twice ten thousand times ten thousand of mounted troops, [74] and they ride creatures like locusts, with tails like scorpions, to sting and torment those who do not have the seal of God on their foreheads. [75] He was worshiped by the Romans as a god—they called him Apollo, and they used the locust as his symbol. He had possessed the emperors Caligula, Nero, and Domitian, perhaps the most evil of Roman emperors, for they all claimed to be incarnations of this great fallen angel. He spoke as in subdued thunder. "Others say he is Francis of Assisi, for he is a stigmata, and bears the marks of Christ on his body, as did Francis."

Satan burned with hate.

"You're fools. Does anybody know? Who is Elijah? The prophet Malachi said Elijah will return before the end times. [76] Is this he? Is the Apocalypse coming?"

Silence.

The Destroyer, the great demon known as Abaddon or Apollyon, spoke again.

"It matters not. He is of no importance. We're winning. We're destroying souls and tormenting mankind in your name."

Satan hissed. "You're fools, and I'm angry. He wears the armor of God, and he speaks with authority. Someone has rescued John Arnold from our snare of money, and someone has rescued his son Matthew from our snare of drugs. It's not a coincidence. You gave John Arnold powers to gain wealth, thinking to poison his soul, but Elijah called upon Jesus to rescue him. Now that wealth may be used against us. It may be used to warn people and to spread—"

Satan stopped. In Hell, the word "Gospel" is not uttered.

"Fear not master," said Beelzebub. "The human realm is under our control. Our greatest trick, our greatest strength, is they think we don't exist. And as long as they think we don't exist, they are defenseless against us. I marvel every moment. More than 7,000 years of torment without end, and they still think we don't exist! THEY THINK WE DON'T EXIST!!!"

At that, the demons roared. And demonic laughter echoed through all of Hell.

The Destroyer spoke with pride. "We are destroying their world and their culture. Suicide, addiction, and mental anguish, deaths of despair, are soaring. Promiscuity is rampant, and sexual perversion grows without bounds. Europe is almost ours. The United States is approaching collapse; the political parties hate each other, and distrust multiplies every election. They don't see it coming, they won't see it coming, they can't see it coming. We are destroying their world. We are working to divide Russia among private armies when Putin falls, and to give them nuclear weapons. The Chinese are enslaved by their government, and we are turning that great country into a military power without conscience, a force of destruction. We are winning. In fashion and art, dark and gothic are the rage. Eurovision parades our disciples. We own most of their universities, they graduate leaders blind to us and blind to God. They can't see God's hand in the universe and life."

Satan sneered. "How's our good friend Darwin?"

The demons laughed again. Beelzebub answered with fake humor. "He is quite well. We keep him chained in our deepest and darkest dungeon and torment him every day. He groans in agony. And Hawking is chained next to him."

Satan burned brighter with hate. "You've done well, but you're fools. Jesus says he will return to defeat us."

Satan paused. "But did God really say that? And will he? Who knows?"

Satan grew larger and burned brighter, and he spoke with terrifying power. "Go forth and spread evil. Capture and torment souls."

Then John saw Satan turn and look at him. Satan's eyes were like lasers. Now Satan grinned and spoke in a deathly whisper, "And bring me the head of John Arnold."

CHAPTER FIFTY-THREE

For the second day in a row, Matthew was up early. He wasn't troubled; there were things on his mind. He felt blessed, and he was trying to take it all in, to make sense of his life, to take stock of where he was, where he wanted to go, and who he wanted to be. He needed to find the real Matthew, the person hidden for years behind the demons of drugs and addiction, the person before prison, before homelessness. He was the prodigal son, lost but now found. He had not just been lost to his parents; he had been lost to himself.

He sat again in the sitting room, outside the second-floor bedroom overlooking the pond, nursing a steaming cup of coffee. The room was spectacular, with fancy woodwork, fancy furniture and fabrics, and expensive art. He wondered how much the art cost, and immediately realized he didn't care. All this fancy, expensive stuff wasn't him. He was different now. He wasn't the spoiled kid who used to take it all for granted—his parent's wealth and all the privileges and opportunities it had afforded him, privileges and opportunities he had received as if he was somehow entitled to them, privileges and opportunities he had betrayed and abused. The room, the house, and the lifestyle were no longer him. Perhaps, he thought, they never really were.

The grace of God was upon him. So much, so very much, to be thankful for. It wasn't just the caffeine. It was an overpowering sense of gratitude. Jesus had given him a second chance, a chance he did not deserve. Jesus had banished his demons. Jesus had given him Imani. Now he had to be worthy.

His father had given him money. A trivial amount for his father, but life-changing for him. A trust paying $50,000 a year if he stayed sober and was working. Life changing.

He had been going to let Imani sleep in, but then he made a decision. And once he decided, he needed to wake her up.

Imani smelled the coffee and opened one eye. "What time is it?"

"Just past seven." "Go back to sleep."

"Please get up. It's important."

Imani grumbled good-naturedly but did as Matthew asked and got dressed, this time in jeans and a sweatshirt. Again, Mathew led her down the back stairs, again out the kitchen to the conservatory. This morning was overcast, but the conservatory was still wonderful and filled with the heady scents of flowers and herbs and the sound of water running down the four levels of the fountain. They were back in Eden.

Again, Matthew led her to Bernini's masterpiece, the 10-foot-high, larger-than-life reproduction of David about to kill Goliath. Time to be worthy. He was nervous but determined.

"Why are we back here?"

Matthew didn't answer. Instead, he dropped to one knee. Imani stared. Suddenly, she was nervous too.

"I know we haven't been together that long," began Matthew. "I know we come from different worlds.

"I've done bad things. I've had opportunities handed to me, and I threw them away. I take responsibility for that, I have fallen short, I hurt my family. But I swear to you, as God is my witness, I am changed. The old Matthew is dead." Matthew stared into Imani's eyes.

"I don't deserve you. I will never deserve you. You are the light that guides my way, you are the part that makes me whole. I love you, and I want to spend my life with you.

"Will you marry me?"

There was a short silence, more than a millisecond but hardly a full breath.

"YES!" Imani pulled Matthew to his feet and hugged him long and hard. She wore tears of joy. Was this really happening? This wonderful man wanted to marry her! Her! A young woman of no background, no money, no education—a simple person. She couldn't believe it. The man who had stolen her heart, the man she was crazy about, wanted to marry her. It was too wonderful.

But then, a thought intruded. She would have to give up her prejudice against money. His family was obscenely wealthy. Then she relaxed. She was tough; she could learn to live with it. But inside, she was still, and always would be, Imani.

She had one request. "Your family wealth is still a shock. I'm simple. I won't have a princess wedding. It's not me."

Matthew took a step back and smiled. "Well, that is an issue. Marriage is compromise. We can negotiate the details."

"You sound like your father."

"The apple doesn't fall far from the tree."

Imani took a step back and laughed. "Alright, but today, we say nothing. If we tell your parents, we might end up with those of those pretentious spectacles. So, promise me, we'll say nothing today. Nothing!" She playfully stamped her foot. "Nothing!"

They walked to the bench on the east side facing the pond and sat talking and laughing. Morning had broken. It was a new day.

Later that afternoon, after saying fond goodbyes, they left for home. Imani had a shift at the hospital and Matthew had dishes to wash. Life was wonderful.

And Imani got her wish. They didn't tell his parents.

At least not then.

CHAPTER FIFTY-FOUR

Mark was getting ready to leave. He'd promised Tiffany he'd get them to Maui, even if late. And why not? He couldn't wait to check out Tiffany's girlfriends and hobnob with their celebrity boyfriends. But first, he needed to talk to his father. Right away, alone if possible—to put Plan C-cret into motion.

His father and mother were downstairs in the kitchen, sitting by the fireplace, drinking coffee and smiling and laughing. There was no time to waste.

"Morning!" greeted Mark, with as much cheerfulness and friendliness as he could muster. "How you feeling, Dad? You look good!"

His father was well and his mother was on top of the world. With the pleasantries out of the way, Mark could get down to business.

"Dad, I've been thinking about what you said. I'm proud that you want to use your money to serve God. That's great. Fantastic. I've been thinking. I've got ideas for corporations and businesses that will help people. We can build homeless shelters and buy farmland in California to feed people. Friends of mine think they can regrow human tissue, including limbs and organs. They need funding. Let's use the money to do those things and more. I'll take the lead and figure out how to change the world for the glory of God. Put me in charge, and I'll do it for you and for God."

Mark was quite pleased with his little speech. That last part had been unusually eloquent. But, annoyingly, his father seemed ambivalent.

Then Mark sensed a small movement. He turned to find his sister standing right behind him listening to every word he'd said.

***

In a way, Ashley wasn't surprised. Mark had always been a jerk. But it hurt. She had hoped she and her brother could put down sibling swords and work together, perhaps even grow to respect each other. But it was not to be. She couldn't yell and scream at Mark, not in front of her father and mother, not right then. She had to play nice. She was going to have to wing it. But confusion clouded her thoughts. Something had changed inside her and was continuing to change. She couldn't quite put her finger on it, but she knew what she wanted from life today was not what she had wanted yesterday.

"Morning, everyone!" said Ashley, with a cheerful and only slightly phony smile. "What's happening? How are we this morning?" She walked up and stood next to Mark. She shot him an "I'll-kill-you-later" glance.

Everyone greeted her back. And her father did look incredibly well, for which Ashley was relieved and grateful.

"Those are interesting ideas, Mark. We should put some thought into them, and other options too. There's so much to consider. Dad, Hajid and I would be happy to help. Not to intrude, but only if you want us to. We have connections, smart people, who can help you with *all* the options. There's no need to rush. You should give this a lot of thought. Take your time, take all the time you want."

Mark shot Ashley a look of disdain. Stalemate.

CHAPTER FIFTY-FIVE

Mark and Tiffany had a car pick them up. The plan was to refuel in San Francisco, then fly on to Maui. The chartered jet was pristine. Mark swiveled contentedly in a soft beige leather chair, flanked by computer screens. They took off a little after 9:00 am, and, at 9:30 am, while they were thousands of feet in the air, the markets opened. Mark stared at the computer screens in horror as crypto collapsed. He watched helplessly as the sell-off accelerated. Three months prior, Mark was worth $100 million. Last week, he was worth $20 million. Now it was gone, all gone.

Mark sat stunned.

"What's wrong?" asked Tiffany.

Mark stared at the screens. His eyes were glued. Finally, he answered, his voice barely a croak. "Crypto is down, down badly, and it's getting worse. It's a disaster." Mark paused. "As of this moment, I am bankrupt."

"Well, let's pray it comes back. You've still got the smart house company. You're smart, you're young, it will work out," said Tiffany brightly.

Mark swung his chair abruptly, and was loud. "And how, *how* is it supposed to work out? I was leveraged. I bet everything on crypto, and the smart house company is in the red. I'm lost."

"Maybe the market will come back."

"You don't get it. You don't understand. You smile and make money. I'm lost. My notes will be called."

"Your father will help you. It will be okay." "Don't be an idiot. He doesn't care. He's giving it all away, remember? His cancer drugs give him hallucinations and now he has to give it all away. He's not going to leave it to his kids like every other father. He's an embarrassment."

"You don't think it was true, what happened to him?" "Of course not!" Mark stood up. "Seriously? I thought you had at least some semblance of a brain. How could anyone fall for that nonsense about a presence?"

Tiffany held her ground. "I believed him. I think it happened."

"Then you're an idiot. I've worked so hard for years and years. I did everything the right way. I studied hard in school. I worked my butt off building companies. And now, I'm bankrupt. It's not fair. I deserve to be rich. I'm bankrupt, and I'm flying in a plane with an idiot."

There wasn't much to say after that. Mark sat down and turned the chair away from Tiffany to face the computer screens. He sat with his back to Tiffany for the remainder of the flight, cursing occasionally, slamming his fist down from time to time.

It took hours, but it seemed forever. They landed in San Francisco and taxied over to refuel. Tiffany pressed the button to talk to the pilot.

"Hi, Jim. I'm getting off here. Please take my suitcase off the plane."

Mark looked at her. "What's this?"

"You heard me. I'm getting off. Where you go from here is up to you."

"I can't believe this!" Mark fumed. "With what I'm going through now, right now, now you want to leave?"

Tiffany said nothing.

Mark lowered his voice and said sadly, "When I need you most, you leave?"

Tiffany reached behind her neck. She took off the diamond necklace and handed it to Mark. "You don't need me, and you don't love me. There's only one person you need, and it isn't me."

Mark was silent. The jet stopped and the door opened. Tiffany walked down the stairs, pulled up the handle on her suitcase, and rolled it towards the terminal.

Mark stood at the stop of the stairs. Tiffany was forty feet away when he shouted. "And who is that, this person you think I need?"

Tiffany stopped, turned around, and gave him one last look.

"Jesus."

She never saw Mark again.

CHAPTER FIFTY-SIX

Some moments define a life. Ashley's moment was just before she joined the conga line. For when she saw her daughter rise with glee, there was an instant, and only an instant, when Ashley prayed for faith. It wasn't a long prayer, and there weren't any words. It was an intense wish for God to enter her soul. And God heard it, and God granted it. In the dimension of the spirit, Jesus forever banished Ashley's demons.

Looking back the next day, months later, years later, Ashley would wonder. Was she just a mother who wanted to have fun with her child? Was it the wine? But deep down she knew, she forever knew, it was a gift from God. Ashley's lens, Ashley's worldview, was forever changed. The curtains of intellectual arrogance, and blind obedience to academic folly, fell.

With her new lens, the glory of God was revealed. She had eyes to see. It was clear, obvious, a no-brainer, that Darwin's theory was a lie. Looking back, she was surprised she ever believed it. It was the stupidity of the mathematics. The theory that keeping the best mistakes—accidental, purposeless, random mutations in DNA code—created the majestic technology of life was revealed as mathematical stupidity of the highest order, a perverted joke of a scientific theory, a lie of the Devil.

Free of demons, Ashley saw that only God could create life. She read continuously when they returned to Boston, book after book and article after article. She was astonished and awed by the complexity and beauty in every cell, the ultra-advanced technology we call life. From then on Ashley would smile every time she saw an article about some scientist who claimed to have found a new clue for the origin of life. She laughed at the things people would say to get their names in the paper.

Stupid mathematics was not Ashley's thing. It wasn't just the perfection of the DNA molecule, with an outer shell protecting code on the inside, and a symmetry that allowed it to be "opened" and copied. It wasn't just the fantastic machinery that read and copied the code, nano-technology orders of magnitude beyond human technology, technology that had to be there at the beginning, technology that could not have created itself. It wasn't just the perfection of the central dogma and other wonders of the process used by life's 3D printers to read the code and build parts.

It was the stupidity of the mathematics. Ashley read that, with each generation, each transfer of DNA coding from human parent to child, there are one hundred mistakes, errors, mutations, more or less. One hundred copying errors. Maybe more. And natural selection,

Darwin's imagined substitute for God, basically gets one choice—the child lives to have its own children, or it does not. The simplicity of the mathematics shocked Ashley. How could anybody have ever thought that Darwin's one choice could defeat one hundred errors? And somehow, magically, mythically, build the majesty of the human brain? It was one hundred errors versus one feeble choice.

Evolution, the change over time of human beings, was down. There was no denying it. Like typos in a book, like rust spots on a car, the errors were adding up.

Hajid didn't have a moment, but it didn't take long. His demons had been weakened by Imani's voice and Elijah's words. They were crushed by the power of Ashley's conversion. Ashley's logic burned with power, and she hit him with it hard.

His demons fought. They threw bolts of arrogance and fear—fear of no longer being part of the club, fear of being different. But Hajid was a man of integrity, and a man seeking truth. With eyes to see, Hajid stepped into the light.

He saw the tension between the first and second laws of thermodynamics. The law of creation—that matter/energy can neither be created nor destroyed—and the law of death—the downward spiral of existence— demanded a God of power and glory, a Creator, and a moment of Creation. The glory of the universe, and the destruction caused by the law of death, demanded a starting Creation of impossible order and perfection. [77] Hajid and Ashley studied and marveled.

They marveled, and they prayed. They joined the church two blocks away. For the first time, they read the Bible. They saw the mystery and wonder of the Christian faith. They thirsted for God as deer thirsts for water. [78] For Ashley, it was joy. Pure mathematicians are cousins to mystics, and faith was a surprise but not a barrier to her career. For Hajid, it was hard. His views clashed with the rulers of academia. His papers were no longer accepted for publication, and his graduate students found other mentors. Then he was summoned to a meeting of the Harvard faculty.

CHAPTER FIFTY-SEVEN

It was a day the Lord had made. Spectacular for anywhere, but for the South Bronx, in the middle of February, more than incredible. Perhaps it was due to climate change or perhaps just pure luck but, on this particular Valentine's Day, the day Imani had chosen for her wedding, it was close to 70 degrees and the sun was blazing.

It came together quickly. Imani's mother had died two years prior, so Ashley, the new Ashley, the Ashley with new eyes to see and a new heart to love, adopted her as a little sister and helped with all the details. Mary was over the moon, of course, and everything she was allowed to do, she did. Her baby, the prodigal son, was now found— and getting married to an amazing woman.

Still, the effect was comical. While most of the guests walked or took the subway, an assortment of chauffeured cars and bodyguards littered the outside of the church. Inside, on the bride's side, it was packed, heavily African American, all laughing, smiling, and spilling over to the back of the groom's side.

The groom's side was schizophrenic. The people in the front rows, mostly John and Mary's friends (although many begged off with excuses), were dressed in expensive clothes and elaborate jewelry. Some wore a nervous look, as if not sure of what was happening and hoping they would get out alive. It was a nervous Caucasian island in a joyful rainbow sea, for behind them, and around them, was the overflow of locals. Everyone in the church had been invited, and everyone who could be there was there.

The choir was in full force, and the music director was beaming. The organist began to play Pachelbel's Canon.

"I'm going to cry," said Mary.

It wasn't much of a procession, but it was perfect. Rebecca walked slowly down the aisle, with flowers in her hair and flowers in her hands. She wore a white lace dress and held her head proudly.

Then the organist began to play a piece from the opera *Lohengrin*, written in 1850 by the German composer Richard Wagner. Eight years later, it was played when Queen Victoria's oldest child married a Prussian prince. It was, of course, "Here Comes the Bride," and all stood and all eyes turned to the back of the church.

Imani walked down the aisle alone. She gave herself away. She had given away her heart to the man she loved, and now she was committing, before God, to join Matthew in a sacred sacrament.

John whispered in Mary's ear. "She's beautiful, and the dress is perfect."

Mary smiled. She had lost that battle. Imani vetoed her offer of custom couture. Imani rented a dress for $45, a simple Black Halo mini, pure white and unadorned perfection. She had changed her diet, lost fourteen pounds, and was now working out. She did accept Mary's offer of a hairstylist and the gift of a simple silver cross to wear around her neck. Gasps and tears welcomed the new Imani into her new life.

Matthew waited at the front, dressed in a new custom suit; he didn't dare turn down his mother's offer on that. He stood in amazement, like a deer in the headlights, stunned by the new Imani.

The soon-to-be newlyweds took their seats on the side of the altar. Mary beamed with pride as Ashley walked up to give the first Bible reading.

"A reading from Paul's first letter to the Corinthians:"

"Love is patient, love is kind. It does not envy, it does not boast, it is not proud. It does not dishonor others, it is not self-seeking, it is not easily angered, it keeps no record of wrongs. Love does not delight in evil but rejoices with the truth. It always protects, always trusts, always hopes, always perseveres.

Love never fails." [79]

The minister walked to the pulpit.

"Dearly Beloved, we are gathered here to witness the marriage of Imani and Matthew. God has put them together, and what God has put together let no man, or woman, tear apart.

"For it is love that breaks down barriers, and love that brings people together. Love will sustain you in the years to come, and love will make your marriage stronger each day.

"This is the marriage of two worlds. But we are not here to list our differences. We are here to celebrate our humanity, and to ask for blessing from our Lord Jesus Christ.

"For Jesus turns the world upside down. Let us listen to Jesus's own words, from his great sermon, the Sermon on the Mount:"

"Blessed are the poor in spirit, for theirs is the kingdom of heaven.

"Blessed are those who mourn, for they shall be comforted.

"Blessed are the meek, for they shall inherit the earth. "Blessed are those who hunger and thirst for righteousness, for they shall be satisfied.

"Blessed are the merciful, for they shall receive mercy. "Blessed are the pure in heart, for they shall see God. "Blessed are the peacemakers, for they shall be called sons of God.

"Blessed are those who are persecuted for righteousness' sake, for theirs is the kingdom of heaven.

"Blessed are you when others revile you and persecute you and utter all kinds of evil against you falsely on my account. Rejoice and be glad, for your reward is great in heaven, for so they persecuted the prophets who were before you." [80]

"What does Jesus mean? What does he mean when he says the poor in spirit are blessed? For Jesus turns the world upside down.

"Jesus tells us to be humble. 'Poor in spirit' means humble in spirit. Not boasting in our righteousness, not claiming superiority over others, not looking down at others. The world tells us to be proud, to get what we can get. The world tells us we deserve it all, and we can do it on our own. But Jesus turns the world upside down.

"Jesus tells us to be humble. It's first on his list. To admit we can't do it on our own, to admit we need God, and to ask for his mercy. To overcome the demons of pride. And if we do that, if we are humble in spirit, we are blessed. And not just blessed, but envied. Envied! Envied, that's what the Greek word means, envied among all. For Jesus turns the world upside down.

"It's the most powerful sermon ever given, the sermon we all need to hear. If we are humble in spirit, the kingdom of Heaven is ours. There is nothing more valuable. It's the pearl beyond price. Jesus turns the world upside down.

"Jesus tells us we are blessed, to be envied, if we mourn. That's not what the world tells us. Jesus turns the world upside down.

"Jesus tells us the meek will inherit the Earth. Outside of Christianity that makes no sense. But Jesus tells us the meek will join him in paradise. Jesus turns the world upside down.

"Hunger and thirst for righteousness, and you will be satisfied. Be merciful, and you will be shown mercy. Be pure in heart, and you will see God. Make peace, and you will be called a child of God. For you see, my friends, Jesus turns the world upside down.

"And if you do these things and you are persecuted, if others utter all kinds of evil things against you, rejoice and be glad. Your reward will be great in Heaven, for so they persecuted the prophets, the martyrs, the believers before you. Jesus turns the world upside down.

"Jesus tells us this world is not all there is. Beyond it is something more valuable. Beyond is the Eden we were made for, the paradise to come, the New Kingdom of God.

Jesus will come again to judge the living and the dead. You will meet Jesus in your lifetime. And in this sermon, this great sermon, the Sermon on the Mount, he tells us how to prepare.

"Jesus turns the world upside down. "Let us pray.

"Heavenly Father, we ask you to bless the marriage of Imani and Matthew. Help them to overcome the obstacles the world will place before them. We know that in this world we will have troubles, but we take heart, for you have overcome the world. Help them and help us to be humble in spirit, to thirst for righteousness, to be pure in heart, to seek peace, and to be merciful. All this we ask in Jesus's name. As he upended the money tables in the temple, as he rose from the dead, Jesus turns the world upside down.

"Amen."

The minister sat down, and the choir stood up. Feet were shuffled, throats were cleared. The rich tones of the organ vibrated the air, and the choir began to sing. They sang "Amazing Grace." Imani and Matthew joined the choir, not as soloists or even lead singers. The whole congregation sang, "Amazing grace, that saved a wretch like me," filling the church with a joyous sound like no other.

The music was followed by communion, the Holy Eucharist, the Christian ceremony replicating the Last Supper, when believers are strengthened by the body and blood of Jesus Christ. All of the church members went to the rail for communion, but only half of the Caucasian island went up.

When communion was over, one of the acolytes brought out a guitar and handed it to Matthew. Mary turned to John in surprise. "What's this?"

"I have no idea."

Matthew walked to the side of the church and stood in front of the choir. Imani walked behind him and stood with the choir.

"I want to sing a song for you," announced Matthew. "Imani and I wrote it. It's a song about my life. It may be about your life too. I'll sing the verses, and Imani and the choir will sing the chorus." He smiled. "The chorus is simple, please join in."

There was no organ. Matthew simply played the guitar and sang the verses, while everyone else joined in with the chorus.

When I was young, I went to school.
Learned it was all about being cool.
They said the smart knew all there was,
But I don't know about that.

I don't know, I don't know.
I don't know, I don't know.
I don't know, I don't know.
I don't know about that.

Got me a job, to earn my way.
They told me I should work night and day.
They told me money was the way,
But I don't know about that.

I don't know, I don't know.
I don't know, I don't know.
I don't know, I don't know.
I don't know about that.

Looked for a woman to share my life.
They told me she must be prim and nice.
They told me she must do what I say.
But I don't know about that.

I don't know, I don't know.
I don't know, I don't know.
I don't know, I don't know.
I don't know about that.

Looked for a reason for my life.
They said don't worry, just be nice.
They said be happy it's all for fun.
But I don't know about that.

I don't know, I don't know.
I don't know, I don't know.
I don't know, I don't know.
I don't know about that.

Walked in a church to hear what they say.
They said that I should bow down and pray.
They said that Jesus Christ was the way.
Now I know about that.

Now I know, now I know.
Now I know, now I know.
Now I know, thank God I know.
Now I know about that.

As his voice trailed away on the last note, Matthew looked out to a sea of smiles and glistening eyes. It was time for vows, so he laid down his guitar and went to stand at the minister's side facing Imani.

The minister asked. "Do you, Matthew, take this woman to be your lawfully wedded wife?"

Matthew was nervous but his voice was clear. "In the name of God, I, Matthew, give myself to you, Imani. I will support and care for you by the grace of God: enduring all things, bearing all things. I will hold and cherish you in the love of Christ: in times of plenty, in times of want. I will honor and love you with the Spirit's help: forsaking all others, as long as we both shall live. This is my solemn vow."

When her turn came, Imani answered in her beautiful voice.

"And I, Imani, in the name of God, take you, Matthew, to be my lawfully wedded husband. I give myself to you, forsaking all others. I will hold and cherish you in the love of Christ, and I will support and care for you by the grace of God. This is my solemn vow."

"I now pronounce you man and wife."

***

Outside the church, invisible demons circled. The bodyguards sensed them and were nervous, but did not know why. The demons cared not about the bodyguards. They had recruited three teenagers who despised God, who were to walk into the church with hidden automatic weapons and slaughter those inside. But the demons snarled in frustration. For the demons could see what the bodyguards could not see, a phalanx of angels protecting the church.

CHAPTER FIFTY-EIGHT

The meeting, if that's the word to describe it, was held in Massachusetts Hall, Harvard's oldest building, in the Perkins conference room on the second floor, looking out over Harvard Yard. Hajid walked in at 9:55 am, five minutes early. He was surprised to find the room full and people already seated. Some of Harvard's most senior professors were there.

The president of Harvard asked him to sit down. There was an open seat to his right at the end of the table. They had been waiting.

The president began. "Professor Akbas, Hajid, thank you for coming. You have brought distinction to this university. But your recent statements threaten the integrity of Harvard, and the reputation of what we and those before us have worked for 380 years to uphold." She paused and looked sternly at Hajid. "Did you actually state in public that the universe is only thousands of years old?"

Hajid scanned the room. These were notable academics, some equal in reputation to him. There were looks of disdain and disbelief. He was on trial, that was for sure.

Hajid cleared his throat. "Yes. I have opened my eyes, and the evidence is overwhelming."

The chair of the physics department hit the table loudly with her fist. She was visibly upset.

"I don't understand," she said, "I really, truly, honestly, don't understand. Your own computer simulation has helped us understand the origin of the universe, from the moment of the Big Bang. You've made a spectacular contribution, one that will be remembered in the history of science. Now, you throw it away and seek to bring us down with you. What happened?"

Hajid looked at her sadly. "It doesn't work. There's so much it doesn't explain, so many facts contradict it."

Hajid caught the eye of the chair of the astronomy department. They had been close friends for over a decade, attended conferences and published articles together.

"The evidence from astronomy is overwhelming. Galaxies have huge numbers of giant blue stars, stars that burn so hot they will burn out, run out of fuel, within a few million years, sometimes in less than a million years. Yet they are everywhere in galaxies we say are billions of years old."

His friend, now perhaps his former friend, looked at him sternly. "But those galaxies are almost unimaginably far away, and light has been traveling from them to us for millions and billions of years. How can you say they are young?"

Hajid paused. "I'll admit. I'm not sure how God did that. There are theories. But galaxies far away fit the Bible. In the Book of Isaiah, and in many other places in the Bible, God tells us he made the Heavens and stretched them out. [81] I've thought about it, but I'm not sure.

Perhaps when we don't know how God did something, we should give the Creator of the universe the benefit of the doubt."

His once friend snorted in derision.

"Look," said Hajid, "your own theories have a problem with distant starlight. The universe looks the same in all directions. You and I have published papers on cosmic microwave background radiation. The galaxies are too far apart for light to have traveled between them and evened out the temperature of this radiation. This evenness, this 'flatness' as we call it, is unexplainable by your theories."

There was another snort from the chair of the astronomy department.

"Come on, Hajid. We've solved that problem, and you know it. The flatness is explained by inflation at the birth of the universe, when gravity was reversed and when the universe grew unimaginably fast in an unimaginably short time."

Hajid chuckled. "Be honest! There's no evidence for inflation, this imaginary force! For every problem it pretends to solve, it manufactures a new one. What force accelerated it? What force slowed it down gracefully enough to obtain our universe? It's never been seen, and nobody knows how it could even exist. It's a made-up theory, like dark matter, dark energy, and dark radiation. The inflation theory has failed three scientific experiments."

His friend, now perhaps his former friend, looked down. He could not refute what Hajid was saying, but neither could he see beyond the prison of his worldview.

Hajid continued. "The irregularities in the rings of Saturn and Neptune, even the existence of the rings, demand a young Solar System, one only thousands of years old. The geological activity of the outer planets and their moons, volcanoes and strong magnetic fields, points to thousands of years. Short-term comets, comets with a cycle of less than 200 years, can't be more than 20,000 years old. Stars in the dwarf galaxies in the Local Group are moving away from each other at more than ten kilometers every second—those galaxies don't look millions of years old. The chaotic orbit of Pluto, the recession of the Moon, binary asteroids, it's all there."

The president of Harvard looked at him. It was not the kindest look. "Alright, Hajid, I'll bite. Remind me what a binary asteroid is."

"Where a tiny asteroid orbits a larger asteroid. About one out of six asteroids in the main belt are binary asteroids. Because of tidal effects, they couldn't have survived millions much less billions of years."

There was no response. The chair of the astronomy department deliberately turned away from Hajid and directed a blank gaze to the center of the table.

Hajid caught the eye of his friend in geology. "Look at the sharp contrasts between layers in the Grand Canyon. There's no evidence of millions of years of erosion. Look at fossilized tree trunks that are partly in one layer and partly in another layer. The size of river deltas, the rate of erosion at Niagara Falls and other places, even the rate of erosion of the continents themselves—they can't be billions of years old; they would have washed into the oceans many times over. Look at the rate of collapse of natural arches, such as in Arches National Park in Utah. Look at the decay of the Earth's magnetic field, with a measured half-life of about 1500 years. It can't be billions of years old." [82]

His friend was sad and looked down. "And carbon dating?"

Hajid got excited. "Yes, carbon dating! Carbon dating is more evidence the Earth is not billions of years old! The carbon-14 isotope, with a half-life of 5,700 years, is found in fossils and even diamonds, diamonds that are nearly impossible to contaminate and are supposed to be billions of years old. Carbon-14 should decay far below our ability to detect it within 100,000 years, and it's in fossils and diamonds that you say are millions and billions of years old!"

His friend in geology stopped looking towards him, turned in his chair, and gazed blankly forward.

Hajid caught the eye of a senior professor of biology. "There's DNA, blood cells, and so much more, in fossils, biological material that couldn't have survived millions of years. Living bacteria have been recovered from salt supposedly 250 million years old. Then there's the decay of all DNA, genetic entropy. The human genome looks young, only thousands of years old."

The biology professor turned away.

Hajid turned to another friend of his, a senior professor in genetics. "There's the genetic evidence of a mitochondrial Eve, a mother of all humanity, who lived just thousands of years ago. Similar data from the decay over time of the Y-chromosome all men carry. And all of the other evidence, from the dispersion after the Tower of Babel by paternal lines, to the amazing similarity of all human genomes, that precisely confirms the book of Genesis."

The genetics professor said nothing and turned away. Hajid sighed. It was lost. He turned to the chair of the history department and was pleased the woman at least looked at him.

"All of world history—Mesopotamia, Egypt, China— begins about 5,000 years ago. Hundreds of cultures around the world have memories of a great flood—the Flood of Noah. The original Hawaiians knew the name of Noah. The ancient Chinese symbol for a large boat was a boat with eight people on it, just as there were eight people on Noah's Ark. We now know the Exodus was a real event. Archeologists have learned where the Jews lived in Egypt, and they've found an ancient Egyptian papyrus, the Ipuwer Papyrus, that confirms the plagues brought on Egypt by God—things like the Nile turning into blood, the sun not shining for three days, and the deaths of Egypt's firstborn. How powerful is a God who can make the Sun cease to shine!"

The chair of the history department turned away. "Thanks for being open-minded," said Hajid sarcastically. "The Pharaoh may have been Amenhotep II, and the Exodus probably took place around 1446 BC. And, this is going to make your day, the Ipuwer Papyrus confirms one of the most amazing facts the Bible tells us about the Exodus, that the Egyptians gave the Hebrews their gold and silver jewelry when the Hebrews left." [83] Now, no one was making eye contact with Hajid.

They stared straight ahead in stony silence.

The silence was awkward and painful, but Hajid could not keep quiet. He had more to say. It burst out; he couldn't help himself: "And you think modern man emerged over 100,000 years ago, and agriculture started 10,000 years ago? You *all* think it took 100,000 years of near starvation for someone to figure out how to put a seed in the ground? People living off the land, totally focused on staying alive, and you *all* think it took 100,000 years to say, 'hey, whadda you know, there's seeds in an apple!' Seriously? That's your peer-reviewed Ivy League theory of how agriculture started?"

At that moment, something inside Hajid snapped. He stood up at the end of the table and shouted: "Oh, come on! How can you not see it? One hundred thousand years of watching plants grow before someone puts a seed in the ground? Wake up!"

Now Hajid was furious. He knew he was right, and he would not back down. He kept standing, and he roared at their blindness.

"So, here's how you think it went down. Two guys, Thud and Crud, are sitting under a tree, minding their own business. Then, suddenly, THWOP! Thud gets hit on the head with an apple! He picks it up and, what the heck, he's hungry right? For 100,000 years, Thud's family has been starving—this is your Harvard peer-reviewed theory, right? So, Thud's *really* hungry, and he takes a *really* big bite. And then—wait for it—for the first time, the first time in 100,000 years of apple eating, he bites into a seed! Go figure! It's epic, right? 'Well, whadda you know,' shouts Thud, '*there's seeds in an apple!*' Yep, that had to be the breakthrough.

"Of course! That *had* to be how it happened. An incredible breakthrough, incredible, and it only took 100,000 years! A discovery so profound it inspires Crud.

As everyone knows, he went on to discover toilet paper. Yes sirree, it wasn't Newton who got hit on the head with the apple, it was Thud!"

There was no response. No one moved. Slowly, now self-consciously, Hajid sat back down.

His once friend, the astronomy professor, spoke up and tried to convince him.

"Come on, Hajid. Please. Look at the Moon. See the craters. Billions of years of bombardments. And on other planets and moons too. How can you explain that?"

Hajid paused. They weren't going to like his answer. "That probably was the Flood of Noah. For a while, God changed the laws of physics, and not just on Earth. The Solar System went berserk. Maybe one planet exploded, maybe some of its pieces hit the Moon, the Earth, and other moons and planets. Many are in the asteroid belt."

There were gasps of shock and plenty of sidelong smirks. His once friend chuckled.

"The laws of physics changed? The laws that no one has ever seen change? Those laws? The laws that everyone knows cannot be violated?"

More smirks. Some snuck glances at Hajid. It was worse than he had feared. He tried again.

"But look at the diffusion rates in zircons. Three independent studies indicate intense amounts of radioactivity during the Flood, which throws all of our dating techniques off. How else do you explain fission tracks in zircon crystals, radiohaloes in granite, and helium in zircons?

"And yes, God can change the laws of physics. He's God. He created them, and he can change them. The laws of physics are not the foundation of the universe, there is something deeper. As proof, look at quantum entanglement, it's an experiment that violates general relativity."

Then there was only silence. After what was only forty seconds, but seemed like forty minutes, the president of

Harvard spoke. "What's God got to do with it? Have you been going to church?"

Hajid sat condemned. "Yes."

The president took a deep breath, then sighed. "I think we've heard enough. Hajid Akbas, you are suspended from all academic duties and privileges. The legal department will decide whether further action is appropriate."

The chair of the physics department now looked at him and spoke softly. "I'm afraid there's a problem with the air filters in your office. For your safety, we're going to move you to the other side of campus."

Hajid tried one last time. "And what about truth?"

The President of Harvard looked at him with contempt. "What is truth?"

Then no one looked at Hajid, and no one spoke to him. They stood and filed out, silently, as if he wasn't there. Hajid was now untouchable. No one would recognize his presence. To Harvard, he no longer existed.

Hajid sat alone. A phrase came to mind, a phrase he had recently learned, a phrase from the first Chapter of Paul's letter to the Romans, and he said it aloud.

"Claiming to be wise, they became fools." [84]

CHAPTER FIFTY-NINE

It was Easter. The season of new birth. Snow white lilies and rainbows of tulips. Trees in blossom. Joy and beauty everywhere, for those with eyes to see.

John and Mary were in New York City, on the Upper West Side. They never went back to the condo on Central Park West. They rented an apartment further west and a little more uptown, a nice one-bedroom with den, clean but not special. No million-dollar art, no wine cellar, no conservatory, not even a crystal decanter. Basic. They had walked to the Easter Service at St. Thomas' Church on Fifth Avenue. Two miles each way, but John had made it, although he'd had to push himself. He was feeling good, if tired. He was at peace and without demons.

They moved three days after Christmas. Mary had insisted John be tested and evaluated by the world's finest doctors. They found no evidence of cancer. There was no medical explanation. The doctors said they had seen this before, there were many cases of miraculous cures that could not be explained by medical science. The power of prayer was beyond understanding.

John suggested they leave the money behind, and Mary had been delighted to agree. John was amazed by that, and more in love than ever. Looking back, he realized Mary never needed the money. Mary didn't define herself by it, and she didn't think she was better than other people because they had it. He knew she was delighted, absolutely and genuinely delighted, to be living with him in a one-bedroom rented apartment. She was happier than she had been in a long time. Quite an amazing woman, that was his Mary.

They didn't have the money, but they had each other. John stopped working. He went from workaholic to relaxed, and they loved it.

They were eating Easter dinner, waiting for the ceremony, the big ceremony. John offered to take Mary out, but she wanted to stay in. She cooked a small turkey breast and made stuffing from a box, with a can of cranberry sauce, green beans that had been microwaved, and mashed potatoes from frozen cubes. John made the gravy. Life was good, real good.

"What about that sermon?" asked John. "Fantastic. Strange. Didn't expect a talk on Hell."

"Heaven and Hell," said John, "two sides of the same coin. God gives us eternal life, and we choose."

"Yes," said Mary. "That was new. She said God is the ultimate gentleman. If we don't want a relationship, he leaves us alone. Like criminals who won't repent, we are kept apart, left to ourselves, put in the prison of our sinful souls. That stunned me, the part about comparing Hell to a prison, where God isolates us for the protection of believers. Never thought of it that way."

"Yes," said John. "She said God doesn't send people to Hell; people choose Hell. They choose whether to spend eternity with God. It's a choice, a choice for eternity, and we make it. And if it wasn't for Jesus, the pure undeserved grace of God, the second chance given by God to all of us, we would all be in Hell, because we are all sinners."

"Yes," said Mary. "She said life is precious, but eternity is a heartbeat away, and we need to prepare now for that last beat. She said the Bible is clear—Jesus doesn't take everyone with him."

That was heavy, so John sat thinking. Then he spoke again. "I liked the part about astronomy and the date Jesus died. She said Jesus died at 3:00 pm on April 3 in 33 AD, using the modern calendar. She said we know the exact date because it was Passover and because when Jesus died, the moon rose covered in blood. The Earth's shadow turned it a dark red—a lunar eclipse. And for three hours before that, while Jesus was on the cross, from noon until 3:00 pm, God stopped the sun from shining. It's in all three of the synoptic Gospels—Matthew, Mark, and Luke—and it's mentioned four times in ancient non- re-ligious books."

John kept talking. "God stopped the sun from shining when Jesus was on the cross. It's too powerful. It grabs me, shakes me, humbles me. We are nothing. God dwells outside our universe with unimaginable power."

Mary spoke. "And Jesus will come again to judge the living and the dead. His first coming was humble, born to a poor family in a small town. He will come again with power and glory."

"We don't know when."

"No," agreed Mary, "we don't know when. When Jesus was on Earth, he didn't know when. He said only the Father knows. The mystery of the Trinity is beyond our understanding."

John looked around the small apartment and gave silent thanks to God, gratitude for existence.

"Then there was the part about Jesus being in Hell from three in the afternoon on Good Friday to 7:00 am the following Sunday. Add it up, that's what she said. Add it up. Forty hours. She said when the Bible uses the number forty, it's usually some sort of test or trial, sometimes a period of judgment or cleansing. When Noah went into the Ark it rained for forty days and forty nights. [85] Moses was on Mount Sinai for forty days when God gave him the Ten Commandments. [86] The Hebrews were confined in the desert for forty years after the Exodus before they were ready to enter the Promised Land. [87] Before he began his public ministry, Jesus was tempted by the Devil for forty days and forty nights. [88] King David, and his son King Solomon, each ruled for forty years. Five times Paul was beaten for his beliefs, and given forty lashes minus one." [89]

"Yes," said Mary. "I've been reading. The Bible uses the number forty over a hundred times."

They sat in contemplation. John was free of demons, but he was anxious. "I'm worried," he said. "I never expected this. Do you think there will be trouble?"

Mary put down her fork and looked at him. "What *did* you expect? You, we, have taken on the Devil. We fight not just against people but against 'powers and principalities.' Paul told us that 2,000 years ago. We fight and we suffer against demons we cannot understand. We have taken on the Devil."

"Seems like that," said John. "Truth doesn't count. Political correctness, saying the right things, fitting in, the good life, that's what it's about. When you think about how some people spend their lives, it's comical. What's that show, "Keeping up with the Kooks?"

Mary laughed. "Yep, that's the show. 'Keeping up with the Kooks.'"

John's thoughts returned to the sermon. "She said we suffer from a 'lack of theological imagination.' That's an interesting phrase, 'lack of theological imagination.' Reminds me of that young man at Christmas, Elijah. He said we live in a fishbowl we call the universe, and we cannot see what lies outside. It's another way of saying we suffer from a lack of theological imagination, but simpler to understand."

"Yes, Elijah. God healed you when Elijah prayed and took your hand. I saw light. In a way, I still can't believe it, and yet I saw it. Yes, 'lack of theological imagination.' Colors and music more beautiful than we can imagine. And she said in the sermon we won't be floating around on clouds playing harps or living out some other stereotype. She said God will give us things to do, a purpose for each according to our talents, as the Bible foreshadows."

"You know," said John, "I got a glimpse in the city, when I was suicidal, sitting in front of the great cathedral. Right after the doctors told me my cancer could not be cured. I got a glimpse of Heaven, a glimpse of beauty beyond description. My guess is something like that, an experience of beauty or peace or wonder beyond description, happens to most people in their lives. But they shrug it off, they tell themselves it was a delusion, and they don't talk about it. They get a glimpse of Heaven but turn back to darkness.

"We never found Elijah, did we?"

"No," said Mary. "He was gone. Vanished. Your detective confirmed he went to MIT and got a perfect score on the Putnam. But he left no trace, not even a digital footprint, after he walked out our front door."

John shook his head. So much he didn't understand, so much he couldn't understand. "Yes," he said, "Elijah. He said we live, we have our existence, in the mind of God. That God *is* existence. That God is all, that God created concepts of quantity, and numbers, and built the universe out of his thoughts. Strange yet wonderful. I don't get it."

Mary grinned. "We're not supposed to *get* it, you lovable dunce. That's the point. It's outside our fishbowl and beyond our understanding. We're just supposed to have faith, and that means trust, trust in the living God."

A wonderful calm came over John. He kept talking. "Pray without ceasing, that's what the Bible tells us to do, [90] and I'm going to try. All our money is but a drop against the trillions of the Devil, but perhaps some will open their eyes and be saved. People don't think the Devil is real, but he is, and he tempts us thousands of times a day. Television, magazines, billboards, and countless internet sites proclaim life is all about money, power, health, good looks, and false gods. A thousand times a day, we are told this or that product, this or that car, this or that house, is what we have to have to be satisfied. It's lies. People spend money they don't have to buy things they don't need to impress people they don't like. It's like Charlie Brown and Lucy with the football. The Devil tempts us, then pulls the football away. The Devil lies."

Mary stood up, walked over, and kissed him on the cheek. "I'm lucky to still have you. Love you, John."

"Love you more." He paused. "I give away all our money, and the cancer disappears. And all the doctors say there's no medical explanation. A miracle. They say they have seen it before, but they can't explain it."

Mary had to ask. "Do you wish we still had the money?"

"No."

Mary smiled and looked into his eyes. "This is like the old days when we were first married. Before the money. Remember our first apartment?"

"The English basement that was cold in the winter?"

"Yes. We slept close for warmth and saved to buy

Ashley a warm bunny sleeper."

"We ate simple meals and went for walks." "Those were the days."

"Great days. We didn't appreciate them. I hope our kids learn to appreciate life."

"Youth is wasted on the young."

John looked at the clock. "We should turn on the TV. The ceremony is about to start."

"Do you regret not going?"

"No, not for a moment. Just as Matthew has to stay away from wine and drugs, I have to stay away from money and power. The demons are powerful, and I can never get close again."

They finished dinner, cleaned up, and turned on the television. The ceremony was about to start, and it was being covered by the major networks.

CHAPTER SIXTY

The networks called it a showdown, a showdown between science and God. Trinity Church was surrounded by barricades and police in riot gear. They faced an angry mob of demonstrators from around the world.

It was the opening ceremony of the Rise Above Foundation. John founded it and gave it all his money, one of the greatest fortunes in the history of the world. He and Mary created it to spread the truth of the Bible. John chose Trinity Church Wall Street, at the intersection of Wall Street and Broadway, in the financial district of lower New York City, for the launch. He had made his money on Wall Street, and that was where he was going to give it away. He and Mary loved Trinity Church. The first church, a simple structure, was built in 1698 and destroyed in the Great Fire of 1776. Until 1869, Trinity Church was the largest building in the United States. It was close to the World Trade Center, had been covered in ash, and played a role in the recovery of New York City from the tragedy of 9/11.

Mary turned on the television. The mob was bigger than the cameras could show. Many waved signs and banners with slogans. "Stand for Science." "Defend Reason." "Science Saves." "Fight the Foundation."

In front of the baying mob, an interview was being conducted. Mary had to look twice and still could hardly believe her eyes. "Oh, my Lord!" she said. "Look! It's our old neighbor! Svetlana!" Mary turned up the volume. Svetlana had a new hairdo—this one was more like a soccer ball than a football.

"What's going on?" asked the reporter. "What can you tell us?"

"It's a cult," said Svetlana into the mic. "A cult, plain and simple. A dangerous cult that wants to undo human progress and drag us back to the stone age. A cult that denies science. That's why we, the National Academy for the Glory of Science, sounded the alarm. And look, just look, at this wonderful response. The world has answered. Science organizations all over the world, and people from top universities all over the world, have come. Together, we stand for reason."

The reporter smiled. "What makes this cult dangerous?"

"So much," said Svetlana, "so much. I don't know where to start. But I can tell you this cult denies Darwin! I'm not making that up. I can tell you this cult denies the Big Bang. It's easy to see why these people should not have this much money. And where that money came from is suspicious, if you ask me. It's too much. We've been talking to the government. The SEC is going to investigate John Arnold's stock sales. We hope the IRS will investigate whether he paid his taxes."

"Is it true you've met the Arnolds?"

"It's true. I've even been to their house. I've seen it with my own eyes. They get drunk and dance around and shout like maniacs. I had to leave. It's a cult, I'm telling you."

The reporter turned away to interview a second person, a short rotund man sporting a pink bowtie.

"Oh, my Lord!" said John. "It's that clown!" John and Mary stared as the second interview began.

"And on my left I have the Very High and Very Right Reverend John Jacob Pierpont Dingledorfer the Fourth, recently chosen to be the Grand Celebrity of the One Size Fits All Church. Congratulations, may I call you 'Fourth'? Let me ask you Reverend Dingledorfer, Fourth, is this fight necessary? Does it have to be?"

"No," said Fourth. "Quite sad. Quite. No need. The best churches have evolved beyond old superstitions and forms of worship. Svetlana is right. But I do want to add that, at the One Size Fits All Church, we welcome people who drink alcohol."

The reporter smiled. "And what do you say about Jesus?"

"You know," said Fourth, "there's no need to worry about that. Personally, I think Jesus was a nice chap. But he's divisive. At the One Size Fits All Church, we don't talk about Jesus. We've evolved. This isn't the stone age. We embrace human achievement. And all are welcome. It's easier than ever. This month, and this month only, we're running a special, 60 percent off. For $19.95 a month, you get full access to our new brewpub and bowling alley, plus...."

The reporter cut him off. "Excuse me, Reverend." The reporter grabbed his ear, the camera zoomed in, and the reporter gave it his best smile. "We now take you inside historic Trinity Church, for the opening ceremony of the Rise Above Foundation!"

CHAPTER SIXTY-ONE

The church was packed. The New York Times had run five front-page articles on the dangers of cults. In each, they said that, because of John's fortune, the Rise Above Foundation would be the most powerful and dangerous cult in history. They interviewed top scientists and celebrities. The message was clear. No rational person should believe the Bible is true. Only a Neanderthal would think that.

As the vastness of John's fortune had come into view, as companies had been sold, as the financial landscape had shifted, the world had been stunned. A reporter at the Times, a man with a reputation for provoking outrage, for inciting anger, decided he could make a story out of it. He was out to sell papers, and he succeeded. The reporter's theme—that religion is a cult, a dangerous cult—made headlines around the world.

They came from around the world to protest. Media-induced anger fueled the mob outside Trinity Church. They were angry, they were loud, and they mocked the audacious claim that the Bible was true. But for some, a few people, a small minority of people, the articles and the coverage had the opposite effect. A few people decided to stand up for God, even if it meant walking through an angry mob. The people inside the church had been shouted at, some spat upon, many threatened. Four of them had been beaten. But they were there. They had run the gauntlet. They stood up for God.

An op-ed piece supporting the Foundation had appeared in the Wall Street Journal. The author said that, if the beginning of the Bible wasn't true, why should the rest be true? Then there were articles and interviews about that, and whether you could believe in both Darwin and the Bible. It was divisive. The vote of the Vestry of Trinity Church, on whether to allow the ceremony to be held at Trinity, was close. It prevailed by one vote.

But it did prevail, who knows why (although rumors circulated of a large financial contribution), and the Church was packed. There were a few minutes to go before the ceremony was to start. The reporter outside the Church needed to cut Fourth off. This gave the reporter inside the church time to explain its design, art, and sculpture, the work of people who dedicated their lives to God. She talked about the Church's rich history. It was founded in 1697. Alexander Hamilton and other early Americans were buried next to it.

The camera scanned the audience. It was diverse. John recognized a few faces from Wall Street, wealthy people he had done deals with. He was surprised they were Christians. He recognized others, even a few lawyers.

That made him smile, who would have thought a lawyer could be a Christian? The African American presence was strong. The Asian American presence was strong. The Hispanic presence was strong. And there were others, a surprising number of others, who looked poor, very poor. John didn't know it, but some of these people had traveled far and spent precious money, some all they had, to stand up for God.

The camera did not show a tall, thin Asian American young man wearing a man bun and a t-shirt with a slight rip in the collar standing at the back. Elijah had returned early from an archeological dig in Israel, where he had been part of a team that found an ancient curse tablet placed by Joshua on Mount Ebal during the conquest. It's precise confirmation of the Bible text, [91] and the undeniable proof that the Hebrews knew how to write—the tablet contained in two places the letters 'YHW'—the Divine name given by God to Moses on Mount Sinai, [92] perhaps best translated as "I am Existence"—centuries before scholars had believed the Hebrews knew how to write, had rocked the world. [93]

The ceremony began. John and Mary moved to the couch. The Rector of Trinity Church began.

"Everyone's so serious," said John. "Matthew's wearing his wedding suit and tie! Ha! Priceless!"

"You made him Codirector." "Look at Imani."

"So sweet. She's beaming." "That was a fun wedding."

"They were *so* nice!" said Mary, "Everybody. I was worried, but they treated us like family. Especially the choir. I still can't believe that choir."

John smiled. "The Chinese food buffet in their parish hall was one of the most fun meals I have ever had. They danced and they sang and we laughed. We're lucky the weather was good because the room was too small for the crowd. But I think our Greenwich and Manhattan friends were shocked. I'm sure they were expecting some over-the-top spectacle, a princess wedding. I will never forget that day."

John looked at Mary. "Speaking of days, is it November? For Imani?"

"Late November. You should remember that date. That's your grandson."

John paused and smiled. "A second grandchild. The miracle of birth. Growing 250,000 new brain cells every minute, with technology we cannot understand."

There are moments when one is overwhelmed by the incomprehensibility of life. For John, this was one of those moments. He sat in silence, staring at the television screen. Finally, he spoke. "You're right. I don't remember ever feeling so good. Amazing grace that saved a wretch like me. I'm more alive than I've ever been. The Creator of the universe doesn't need our help. God doesn't need our money. But the Rise Above Foundation can reflect his glory. Perhaps it can give some people, a few more people, eyes to see the living God."

"People are without excuse," said Mary. "Paul said it 2,000 years ago, in the first chapter of his letter of encouragement to the Christians in Rome. 'For his invisible attributes, namely, his eternal power and divine nature, have been clearly perceived, ever since the creation of the world, in the things that have been made. So they are without excuse'." [94]

John asked, "Did you see that article about the James Webb telescope?"

Mary shook her head. "No."

"It's that new telescope in space, the one that's more sensitive to infrared light so it can see more distant galaxies. It's revealing the opposite of what the Big Bang predicts. It can detect very distant galaxies, galaxies that, according to their models of looking back in their imagined deep time, should be shapeless blobs. But that's not what the telescope is seeing. The distant galaxies are large and fully formed. And this finding contradicts, actually destroys, the Big Bang theory, but of course, the big-shot academics can't admit it. The galaxies are fully formed." [95]

John paused. "People are without excuse. The Bible tells us in many places God made the Heavens and stretched them out."

Mary had to ask. "How are they going to hold on to their theory?"

"Who knows? I've been talking to Hajid. He's actually pretty smart."

"Well, well!" Mary gave John a smirk. "Now isn't that a surprise! Can you believe you used to be mean to him? I'm glad you two are friends now."

"I like him," smiled John. "Hajid says probably one of two ways. They may try to backdate their imagined deep time even farther and claim these distant galaxies are really old. Or they could invent another lie. He says they invented this concept they call 'inflation,' with no evidence of course, to try to explain how the universe looks the same in all directions. He tells me they've also invented concepts they call dark matter, dark energy, and dark radiation—three other invented theories to jam what they are seeing into their glass-slipper model of the Big Bang. He thinks they'll come up with something else and call it 'dark' to hide that they don't have a clue."

John looked at Mary. "How about the Dark Galaxy Shape Shifter Force?"

Mary laughed. "May the force be with you. Did you see they found skin, I'm not kidding, skin, from a duck-billed dinosaur that supposedly died 70 million years ago? [96] The article said scientists are trying to figure out why the skin didn't decay. How about the 'dark hole' theory? Dark holes form around fossils so they don't decay."

John laughed. "People are without excuse. I saw an article, it was even in the Times, about octopuses that build shells. Did you know octopuses have eight arms with suckers, three hearts, and other stuff that's even weirder? They are wicked smart; they can open jars and navigate mazes. The DNA of an octopus is so different from the DNA of any other creature that some scientists have suggested it came from outer space. I'm not making that up. It was signed by over 30 big-shot scientists and published in a major journal. [97] But this new article wasn't about that, it was about the DNA of a kind of octopus that builds shells. They said the shell-building coding, the shell-building DNA, is totally different from the shell-building coding/DNA of crabs and clams. [98] They were shocked, to say the least."

"I hate to ask," said Mary, sarcastically. "Where did they say that new coding came from?"

"Ha!" said John. "It 'evolved.' That's all the article said. 'It evolved.' I'm not kidding, all it said was 'it evolved.' No discussion of how useless, stupid, and mathematically absurd that statement is."

John paused in thought, then turned back to Mary. "I've been talking to Ashley. She is amazed at how mathematically bankrupt Darwin's theory is, now that we know all life is built from coding, the coding of DNA. She's tried to explain it to me, not sure I understand. I think one of her examples was we have four million switches in our body to turn our systems on and off. This is a published finding by over 400 scientists, I'm not kidding. We have four million switches, and each is an engineering marvel. [99] She says the suggestion that these four million switches were all created by mistakes at the same time is, mathematically, nonsense of a high order, and equally stupid is the suggestion that, even if somehow one of these switches is created by accident, it remains unchanged while the other millions 'evolve.'

"But, of course, the article didn't mention that. Not a word. It would be funny if it wasn't so tragic."

John turned his attention back to the TV. The Rector finished his welcome. Matthew appeared and said a few words. Then it was Ashley's turn.

CHAPTER SIXTY-TWO

Welcome," said Ashley, spreading her arms wide. "This is a day the Lord has made." Ashley saw bursting pews and people standing in the back. She saw television cameras and heard demonstrators outside. She spoke loudly to be heard.

"Let us begin," she said, "with three questions. What is the Christian faith? What does it hope for? What is the Rise Above Foundation?"

"For a Christian, faith is trust. We trust in the living God. We do not trust in ourselves. We do not trust in our institutions.

"We believe the Bible is true. In the beginning, God created the Heavens and the Earth. [100] God so loved the world that he gave his only begotten Son, that all who believe in him should not perish but have eternal life. [101]

"We believe in Jesus, the only begotten Son of God. For our sake he was crucified, he suffered death and was buried, and rose again on the third day. He will come again in glory to judge the living and the dead, and his kingdom will have no end.

"We look to the resurrection of the dead, and the life of the world to come. [102]

"That is our creed. We believe the Bible is true. That is the Christian faith, the Christian trust.

"The Christian hope is simple. We hope for Jesus to come again, to take us with him to paradise, to a new Eden. As Mother Teresa prayed, 'Come Lord Jesus, come quickly.'"

Ashley paused to take a sip of water.

"And now we ask, what is the Rise Above Foundation?"

Ashley looked around. The outside world was shouting. She continued.

"Perhaps there was a day, perhaps there was a time, when a person of conscience might have cause to deny God. Perhaps there was such a day, before we knew of the wondrous technology of life, technology far beyond anything human beings have made or conceived. Before we knew the universe was created and designed for life. Before we had computers to simulate the downward spiral of mutations and expose the lie that errors and death can create. Perhaps there was such a day, but I doubt it. For it was well-said, close to two thousand years ago, that the divine nature of God, his eternal power and glory, is revealed in the things he has made, and people are without excuse.

"But even if there was such a day, that day is no longer. Even if there was such a time, that time has passed.

True science, the power of two thousand years of thinking God's thoughts after him, two thousand years of searching for truth, has revealed and continues to reveal, in greater glory with each passing year, the majesty of God. We are humbled by the technology in every creature, in every cell. We know Darwin's imagined "intermediate varieties" never existed, they are nowhere in the fossil record, regardless of how many deluded museums invent displays of supposed evolution. And today we know—evolution is down, not up. It is simple mathematics. People are without excuse.

"The theory that death–the so-called survival of the fittest–a nonsensical tautology if ever there was one, can create the technology of life is absurd, a perverted joke of a scientific theory. It has no experimental support. Yet this lie, this lie that goes by the name of Darwin, has led billions to their doom.

"Today, we rise above. My parents—John and Mary Arnold—were blessed with treasure. Today, they store up that treasure in Heaven. For godliness with contentment is great gain. We bring nothing into the world, and we take nothing out. Those who desire to be rich fall into temptation, into a snare, into many senseless and harmful desires that plunge people into ruin and destruction. For the love of money is a root of all kinds of evil. It is through this craving that some have wandered away from the faith and pierced themselves with many pangs." [103]

Ashley smiled. "Those are not my words. Those are the words of Saint Paul from two thousand years ago. They are as true today as when they were written. Maybe truer.

"Jesus said it plainly—where your treasure is, there your heart is also. [104] Today, my parents store up treasure in heaven, where thieves do not break in and moths do not destroy. [105] Today, they pledge their treasure to fight the lies of the Devil. That is the Rise Above Foundation."

Ashley raised her eyes to Heaven.

"It is not, and never will be, an easy fight. We fight against fallen angels, demons that twist and corrupt. We fight against powers and principalities we cannot see. [106] We fight against monstrous lies that cruelly and cowardly lurk behind a mask of science. Today we declare war on that twisted bag of lies. Others may swallow the lie that science is contrary to God. Others may write articles about Darwin's imagined power of natural selection. It does not matter where those lies are written, whether in books or carved in stone, they are lies of the Devil.

"Today, we rise above. We rise above the lie that life is ordinary, and its progeny of bigotry, persecution, and hatred. We rise above the lie that anything can exist without God.

"These lies are spread by many of the institutions of our world. Institutions we have trusted have lied to us. My husband and I have decided to leave the institutions that currently employ us. A great man once said, "Live not by lies." My husband and I will not live our lives in a lie. It is to the great and everlasting shame of many of our most powerful institutions that they have embraced these lies. They do not deserve our respect. They do not deserve our trust.

"We are Christians. We trust in the living God. He is Existence, in him we live and move and have our being.

"I teach math. In math, when you say one answer is closer to correct than another, that means there is a correct answer. It's the same in life. When you say one moral standard is better than another, that means there is a perfect moral standard. God is that standard. Without God, morality crumbles, and power rules. We need God. He *is* Existence. In him we live and move and have our being.

"Amen."

CHAPTER SIXTY-THREE

John could not believe his daughter had delivered so magnificent a speech. Mary turned to him. "Ashley quoted Solzhenitsyn. 'Live not by lies.' That's powerful. She's saying she and Hajid didn't want to work at places that spread the lies of the Devil. I'm still amazed they will leave their schools."

"She's brave."

"Hajid is braver," said Mary. "He has renounced the lies taught by the Harvard physics department. He will never get another grant. Atheism is the official religion of the United States, at least when it comes to handing out research money."

John smiled. "I hope his new book gets traction. He told me dozens of astronomical facts, undeniable facts, point to God and the truth of the Bible. He told me quantum physics proves the universe is mental, not physical, created by the thoughts of God. He's going to write a book. We talked about titles. I suggested 'Let those who have eyes see.'"

Mary looked at him with a secretive grin. "Did you know Ashley is pregnant?"

"No! Are you kidding? Wow! That's wonderful!"

"Yes. I wasn't supposed to tell you, at least for a few weeks, but I can't keep it a secret. Rebecca may finally get a baby brother. Or a baby sister."

John's face fell. "Mark didn't come, did he?"

"No. He's angry we didn't leave him money. He told Ashley she's an idiot to leave MIT. Said he wants nothing to do with her and Matthew."

John paused and looked away. "So sad. He worships money like I used to. He defines himself by money. He cannot break his chains."

"Imani spoke to Tiffany, though."

"Really?" John turned back to Mary. "Haven't heard that name in a while. How is Tiffany?"

"She's taking a break from modeling. She told Imani the trip to Connecticut changed her life, and now everything is different. She told Imani she has given her brokenness to Jesus and is trying to be worthy of his righteousness, the great and mysterious exchange of Christianity. She even said she is thinking of becoming a minister. And she cut her long blond hair."

"Whoa! She is going to be one good-looking minister. People will fight to sit in the front pew."

And so it went. For a few minutes, John and Mary laughed and loved. They had no idea what was about to happen.

CHAPTER SIXTY-FOUR

John could not believe how loud the demonstrators were. It sounded as though they were right outside his window. Then he realized they were.

He walked across the room to look. He was shocked to see demonstrators arriving by the dozens with signs and banners waving. They were loud and angry, and getting angrier by the second, it seemed. Before John could process what was happening, his cell phone vibrated.

John answered and listened intently. "Thanks," he said in a subdued voice and immediately turned to Mary. "That was Luke. He's still looking out for us. He says we have to get out of here now. This address has been posted on social media. There are death threats, and Luke says they are serious."

They took three minutes to pack. John grabbed the book of memories Mary had given him for Christmas and made sure she was wearing the heart-shaped diamond he gave her. They threw a few clothes in a suitcase, grabbed toothbrushes and other items, and were off. With one suitcase and a set of car keys, they headed for the elevator, and down to the garage in the basement.

The keys operated an ancient brown Datsun covered with rust, with a bad dent on the left side and a right bumper that sagged. Matthew's employers were visiting relatives and had asked if they could leave their car in the garage below the apartment, where it would be safer than on the street. John and Mary drove out of the garage in the world's ugliest car.

They hoped the demonstrators would be looking for an expensive car, and they could slip out unnoticed. It almost worked. But when they turned into the street one man recognized them. It was as if he knew exactly where they would be, and when they would emerge. He put down his backpack, pulled out a heavy revolver, and took aim at John's head. They didn't know it, but he was a veteran who had been badly wounded, mentally and physically. He had been trained as a sniper. He was a member of a cult that worshipped Satan. His body and his soul now belonged to a demon. He was close. As John and Mary drove away, he lifted the gun with both hands, took careful aim, and fired.

CHAPTER SIXTY-FIVE

The bullet missed. Who knows why. It pierced the back window, missed John's right ear by one inch, and made a larger hole in the front window. Also surprising was the large truck that immediately moved into the lane behind John and Mary. The gunman did not get a second shot.

They drove in silence. There were no words. Mary didn't know how to feel or what to say. John was trying to process. They drove north to the George Washington Bridge and crossed the Hudson to New Jersey. Cell phones rang but Mary turned them off. They drove south on the New Jersey Turnpike, then took the Pennsylvania Turnpike west. Two hours went by. Neither spoke.

John drove into the setting sun. The western sky exploded with pinks and aquas of incomprehensible beauty. Then the sky darkened and a star appeared. Mary turned on her phone and opened the app Rebecca had installed. She pointed the phone at the star. "It's Jupiter."

John finally spoke. "We follow the star."

He took the next exit to get gas. The self-service station was not crowded. A few people noticed the bullet hole in the back window, and the larger hole in the front window, but no one spoke to them. A chain restaurant was down the street. John used to like their fried chicken, so he parked in the back, far from the front door. After they were shown to a booth, Mary ordered water and John a diet coke.

Mary took a deep breath and managed a weak smile. "This is a fine mess you've gotten us into."

John chuckled. The shock had finally worn off. "A very fine mess indeed."

"What now?"

John's mood brightened. "Well, I've been thinking. How about we keep going? You and me together, leaving it all behind, at least for a few weeks, maybe a few months, who knows? We can't go back to New York or Connecticut, not right now. Could be dangerous."

John looked at his wife of forty years. "What do you think? An adventure? Leave our old lives behind? Keep going? Follow the star?"

John saw Mary think it over, then smile. "That would be wonderful. I would love that! An adventure! A trip like we used to take when we were first married, before the kids and before the money. But on one condition."

John was worried. He looked at Mary. "What's that?"
"No camping."

John laughed. "Okay, you win."

John was excited. "Let's do it! There's a motel on the corner and a car dealer down the street. Tomorrow we buy a car and head out. Something easy on my back."

"Can we afford that?"

John chuckled. "I know a guy who knows a guy. Have faith."

"And the brown Datsun?"

"It's had a good life. I'll arrange a new car for the Chinese couple. They deserve it."

"Where do we go?"

"How about Texas? There's a museum that showcases the truth of the Bible, the truth of Biblical creation. They call it the Discovery Center; it's outside of Dallas. I read the planetarium is life-changing, and they have a dinosaur fossil that is rubbery. I want to see that fossil. I want to touch it. That would be like putting my hand on the mark on the side of Jesus. To touch soft tissue that is supposed to be millions of years old would prove deep time is a lie."

"Then what? Come back?"

"Maybe keep going. I'm thinking the Grand Canyon. I want to see the layers laid down by the stages of the great flood of Noah, 5,300 years ago. Overwhelming proof of the truth of the Bible, plainly visible yet not seen. Some of those layers cover most of North America. I want to see with my own eyes the sharp lines separating the layers, see the lack of erosion and know that millions of years is a lie. I want to see it with my own eyes."

Mary chuckled. "I'd like that. Can we play country music? 'All my Ex's live in Texas.'"

John smiled. "'Tequila makes her clothes fall off.'" "Oooh, frisky!" Mary looked at him mischievously.

"'Want you to love me like my dog.'" John smiled. "You win again."

"Well," said Mary. "Let's check the scorecard. Money?"

John smiled. "Gone. Compared to a few months ago, basically nothing."

"Fancy house?" "Gone." "Wine cellar?" "Gone."

"Armored Escalade and silly cars like that Lamborghini?"

"Long gone." "What's left?"

John took Mary's hand. His eyes sparkled, and he held her hand gently. "I've got you babe. You and Jesus. I've never been so blessed."

John kept holding Mary's hand. "But let's pray, because we need help. We have taken on the Devil, and we need help."

John closed his eyes and prayed:

"Saint Michael the Archangel defend us in battle, be our protection against the wickedness and snares of the Devil; may God rebuke him, we humbly pray; and do thou, O Prince of the heavenly host, by the power of God, cast into hell Satan and all the evil spirits who prowl through the world seeking the ruin of souls."

"Amen."

CHAPTER SIXTY-SIX

At the invocation of his name, the great Archangel Michael took notice. He looked down with compassion. He was in Heaven, an alternate space/time reality created by God. But time in Heaven is not the same as our time. Words cannot describe the difference.

But, to the extent words can describe the Archangel Michael, he was, he is, he always will be, magnificent. He commands the army of God against Satan. The Bible tells us that, in the time of the end, he will arise to fight Satan, [107] and, with the help of other angels, he will defeat him. [108]

The great Archangel Michael spoke to the angels around him. He looked at one.

"May God reward you for protecting John and Mary. That shooter almost killed John."

In reply, the angel spoke softly. "God was with me. I asked him to bend the path of the bullet, and he did."

"It's not over," said Michael. "Satan will pursue John and Mary. We must always be on guard. Satan's forces multiply, and his demons grow in power. The children of God are sorely oppressed."

Another angel spoke softly. "Some fear we are losing. They say God is dead, they mock his creation, and they turn from the light to darkness, because their deeds are evil. [109] What are we to do?"

"Have faith," said Michael. "Faith in God. Pray without ceasing. Give glory to God, and trust in his mercy. For in him we live and move and have our being."

ABOUT THE AUTHORS

Counting to God author DOUGLAS ELL, a former atheist, said "I've spent more than 30 years reconciling science and God, because I needed scientific evidence to believe in God." Douglas Ell graduated early from MIT, where he double majored in math and physics. He then obtained a master's in theoretical mathematics from the University of Maryland. After graduating from law school, magna cum laude, he became a prominent attorney. Ell drafted the first 401(k) plan in professional sports and has represented a number of nationally recognized corporations, unions, and pension plans. His legal training and work, combined with his academic science background and a lifetime of independent study, have given him a uniquely grounded approach to science, religion, and philosophy and the ability to untangle the confused and often emotional relationship between science and religion.

S TEVE EGGLESTON is a law school Valedictorian, former law professor, author, lecturer, and colourful trial lawyer. In his renaissance life, he has launched a hip-hop start-up, produced feature films, helmed a rock 'n roll magazine, booked 1000+ live shows worldwide, and managed Grammy-winning artists. As an international best-selling author, he is published in fiction and non- fiction and lives with his family in Somerset, England, where he draws and paints in his free time. Steve's business can be found on steveegglestonwrites.com.

BIBLIOGRAPHY

[1] Mark 8:36: Unless otherwise noted, all Bible references are to the English Standard Version.

[2] Compare 2 Corinthians 12:4.

[3] Psalm 23:4.

[4] 1 Samuel 17:11.

[5] 1 Samuel 17:26.

[6] 2 Corinthians 4:16–18.

[7] This philosophy is often called "moralistic therapeutic deism." It is counterfeit Christianity, where God places no demands on people. It has become the most popular worldview in U.S. culture. See https://www.arizonachristian.edu/2021/04/27/counterfeit-christianity-moralistic-therapeutic-deism-most-popular- worldview-in-u-s-culture/ (accessed June 22, 2023).

[8] Psalm 19:1.

[9] Here's a short video with John Lennox on this subject: https://www.youtube.com/watch?v=h3DO3PAFFX8 (accessed June 22, 2023). A quick search of the internet will find many more.

[10] Not a real award.

[11] Three well-known scientists have proven that, even if the multiverse exists, the past can't be infinite. It's called the Borde-Guth-Villekin Singularity Theorem of 2006. Alexander Villekin wrote: "It is said that an argument is what convinces reasonable men and a proof is what it takes to convince even an unreasonable man. With the proof now in place, cosmologists

can no longer hide behind the possibility of a past eternal universe. There is no escape, they have to face the problem of a cosmic beginning." Alexander Villekin, Many Worlds in One, p. 176 (2006).

[12] 1 Samuel 17:46–47.

[13] One of God's special requirements for the kings of Israel was that they would hand-copy Scripture. "When he takes the throne of his kingdom, he is to write for himself on a scroll a copy of this law, taken from that of the Levitical priests. It is to be with him, and he is to read it all the days of his life so that he may learn to revere the Lord his God and follow carefully all the words of this law and these decrees and not consider himself better than his fellow Israelites and turn from the law to the right or to the left. Then he and his descendants will reign a long time over his kingdom in Israel." Deuteronomy 17:18–20.

[14] Isaiah 11:1.

[15] Luke 3:23–32; Matthew 1:6–16.

[16] Isaiah 9:2.

[17] Isaiah 9:6.

[18] Micah 5:1–2.

[19] Luke 2: 8–15.

[20] John 1:1.

[21] Compare this also to the vision in Revelation 12:1– 2: "And a great sign appeared in heaven: a woman clothed with the sun, with the moon under her feet, and on her head a crown of twelve stars. She was pregnant and was crying out in birth pains and the agony of giving birth."

[22] Leviticus 1:10: "If his gift for a burnt offering is from the flock, from the sheep or goats, he shall bring a male without blemish…"

[23] There is disagreement over where Jesus was born. For support of Imani's view, see https://medium.com/mormon-writers/

the-true-birthplace-of-christ-the-tower-of-the-flock-
97d02dcbb885 (accessed June 28, 2023).

[24] Isaiah foretold, 700 years earlier, that the Magi would arrive
with hundreds of camels. He also foretold two of the three gifts
they would bring. From the sixth verse of the 60th chapter:

A multitude of camels shall cover you,
the young camels of Midian and Ephah;
all those from Sheba shall come.

They shall bring gold and frankincense,
and shall bring good news, the praises of the LORD.

[25] Isaiah 60:6.

[26] If the dates in this book are correct, and Jesus was born on June
17 2 BC and died on April 3, 33 AD, then he was 33 years old
when he was crucified. There was no year zero.

[27] Philippians 2:10.

[28] 2 Corinthians 4:8–10

[29] This is true. MIT graduations are held in a space called "Killion
Court" facing MIT's main building. Above the columns, prom-
inently displayed in letters that appear to be about forty inches
high, are the names of scientists, written in capital letters. MIT
students receive their degrees beneath the name of DARWIN.
God is rarely allowed in the building.

[30] Matthew 27:65.

[31] 1 Corinthians 15:8.

[32] 1 Corinthians 15:6.

[33] John 1:1.

[34] Acts 17:28.

[35] John 3:16.

[36] Job 19:25.

[37] John 6:40, 44.

[38] John 16:25–33.

[39] Here are Jesus's words to Nicodemus, a Jewish leader and holy man: "Jesus answered, "Truly, truly, I say to you, unless one is born of water and the Spirit, he cannot enter the kingdom of God. That which is born of the flesh is flesh, and that which is born of the Spirit is spirit. Do not marvel that I said to you, 'You must be born again.'" John 3:5–8.

[40] Here are Saint Paul's words to the people of Athens: 'Yet he is actually not far from each one of us, for in him we live and move and have our being.' Acts 17:27–28.

[41] John 1:1.

[42] Here are Jesus's words as recorded in John 16:33: "I have said these things to you, that in me you may have peace. In the world you will have tribulation. But take heart; I have overcome the world."

[43] Job 38:4–5.

[44] See Romans 5:3–5.

[45] Genesis 1:26.

[46] Romans 10:9.

[47] Isaiah 53:5.

[48] John 15.13.

[49] 2 Corinthians 5:21.

[50] Genesis 9:14.

[51] This is a real test, officially called the William Lowell Putnam Mathematical Competition. In the nine years between 2013 and 2022, MIT won seven times, came in second once, and fourth the other year. In 2021 and 2022, MIT students had all five of the top scores, and the overwhelming majority of the top 100 scores were earned by MIT students.

[52] "The human brain sees the world as an 11- dimensional multiverse," New York Post June 13, 2017. For the original scientific article in *Frontiers in Computational Neuroscience* see https://

www.frontiersin.org/articles/10.3389/fncom.2017. 00048/full (accessed July 17, 2023).

[53] *See, e.g.,* Darwin's Doubt by Stephen C. Meyer (2013).

[54] Romans 8:22.

[55] 1 Thessalonians 5:16–18.

[56] Romans 1:22.

[57] A technical discussion of this subject can be found in Chapter 6 of Darwin's Doubt by Stephen C. Meyer.

[58] https://www.smithsonianmag.com/smart-news/neander-thals-hunted-and-butchered-massive-elephants-125000-years-ago-180981578/ (accessed August 13, 2023).

[59] A good book on these incredible extinct animals (with great illustrations) is Untold Secrets of Planet Earth, Monumental Monsters by Vance Nelson (2017).

[60] This dating comes from the Septuagint version of the book of Genesis.

[61] Due to changes in the tilt of the Earth over thousands of years, this shaft now points to a slightly different place.

[62] For an introduction to this complex subject see *MT, SP, or LXX? Deciphering a Chronological and Textual Conundrum in Genesis 5,* by Henry B. Smith Jr., Bible and Spade Winter 2019, p. 18–27.

[63] See https://www.draconika.com/cultures/roman- dragons.

[64] John 1:5.

[65] Luke 2:24.

[66] Ezekiel 28:13.

[67] Ezekiel 28:11.

[68] Ezekiel 28:14.

[69] Ezekiel 28: 15, 17.

[70] 1 Kings 18:40.

[71] 1 Kings 18:27.

[72] 1 Kings 19:1–3.

[73] Revelation 9:11.

[74] Revelation 9:16.

[75] Revelation 9:3–6.

[76] Malachi 4:5–6.

[77] See *Counting To God, A Personal Journey Through Science To Belief*, by Douglas Ell, pp. 79–81 (Attitude Media 2014).

[78] See Psalm 42:1. "As the deer pants for streams of water, so my soul pants for you, O God."

[79] 1 Corinthians 13:4–8.

[80] Matthew 5:3–11.

[81] Job 9:8; Psalm 104:2; Isaiah 40:22; Isaiah 42:5; Isaiah 44:24; Isaiah 45:12; Isaiah 48:13; Isaiah 51:13; Jeremiah 51:15; Zechariah 12:1.

[82] There is overwhelming evidence for the truth of the Bible, including Biblical Creation thousands of years ago, not billions. A good starter article is *Evidence for a Young World* by Dr. Russell Humphreys   https://answersingenesis.org/astronomy/age-of-the- universe/evidence-for-a-young-world/ (accessed July 14, 2023).

[83] Exodus 12:35–36.

[84] Romans 1:22.

[85] Genesis 7:4, 12, 17.

[86] Exodus 24:18, 34:28.

[87] Numbers 14:33–34.

[88] Matthew 4:1-2, Mark 1:13, Luke 4:2.

[89] 2 Corinthians 11:24.

[90]   1 Thessalonians 5:17.

[91]   See Joshua 8:30–34.

[92] See Exodus 3:14.

[93] This is an actual archeological discovery of the first magnitude. The curse table was discovered in December 2019, but it took two years of scientific research and x- ray images to read the Hebrew letters etched more than 3,000 years ago into the center of the folded lead curse tablet. See *Bible and Spade* Spring 2023 Vol. 36 No. 2.

[94] Romans 1:20.

[95] *See, e.g.,* "James Webb Telescope vs. the Big Bang," *Acts & Facts*, Vol. 51, No. 7, p. 14.

[96] See https://answersingenesis.org/fossils/duck-billed-dinosaur-skin-preserved/ (accessed July 13, 2023).

[97] See Newsweek May 17, 2018, https://www.newsweek.com/alien-octopuses-outer- space-930942 (accessed July 14, 2023)

[98] It Looks Like a Shell, but an Octopus and 40,000 Eggs Live Inside", New York Times Nov. 5, 2022, https://www.nytimes.com/2022/11/05/science/octopus-shell-argonaut.html (accessed July 13, 2021).

[99] New York Times, Sept. 5, 2012: "The human genome is packed with at least four million gene switches that reside in bits of DNA that once were dismissed as 'junk' but that turn out to play critical roles in controlling how cells, organs and other tissues behave."

[100] Genesis 1:1. [101] John 3:16.

[102] These sentences are from the Nicene Creed, a fundamental statement of the Christian faith. See https://www.episcopalchurch.org/what-we- believe/creeds/ (accessed July 17, 2023).

[103] Compare to 1 Timothy 6:6-19.

[104] Matthew 6:21.

[105] Matthew 6:19.

[106] For a riveting and true story of one man's encounters with demons and exorcisms, read *Demonic Foes* by Richard Gallagher
M.D.

[107] Daniel 12:1.

[108] See Revelation 12:7. [109] John 3:19.

www.ingramcontent.com/pod-product-compliance
Lightning Source LLC
Chambersburg PA
CBHW071355150726
48000CB00001B/28